Business and Environmental Risks

Diego A. Vazquez-Brust · José A. Plaza-Úbeda ·
Jerónimo de Burgos-Jiménez · Claudia E. Natenzon
Editors

Business and Environmental Risks

Spatial Interactions Between Environmental
Hazards and Social Vulnerabilities
in Ibero-America

Foreword by Kjell-Erik Bugge

Editors
Dr. Diego A. Vazquez-Brust
The Centre for Business Relationships,
 Accountability, Sustainability
 and Society (BRASS)
Cardiff University
55, Park Place
Cardiff
Wales
CF10 3AT
U.K.
VazquezD@cardiff.ac.uk

Dr. José A. Plaza-Úbeda
Department of Business Administration
University of Almeria, Spain
Ctra. Sacramento s/n La Cañada
 de S. Urbano
04120 Almería
Spain
japlaza@ual.es

Dr. Jerónimo de Burgos-Jiménez
Department of Business Administration
University of Almeria, Spain
Ctra. Sacramento s/n La Cañada
 de S. Urbano
04120 Almería
Spain
jburgos@ual.es

Prof. Claudia E. Natenzon
Institute of Geography "Romualdo
 Ardissone"
Faculty of Philosophy and Letters
University of Buenos Aires
Puan 480 - 4° piso
Buenos Aires 1406
Argentina
natenzon@filo.uba.ar

ISBN 978-94-007-2741-0   e-ISBN 978-94-007-2742-7
DOI 10.1007/978-94-007-2742-7
Springer Dordrecht Heidelberg London New York

Library of Congress Control Number: 2011941278

© Springer Science+Business Media B.V. 2012
No part of this work may be reproduced, stored in a retrieval system, or transmitted in any form or by any means, electronic, mechanical, photocopying, microfilming, recording or otherwise, without written permission from the Publisher, with the exception of any material supplied specifically for the purpose of being entered and executed on a computer system, for exclusive use by the purchaser of the work.

Printed on acid-free paper

Springer is part of Springer Science+Business Media (www.springer.com)

To all those living in 'hot-spots' and suffering the consequences of the problems outlined in this book. We wish them a brighter future both socially and environmentally

To our families and friends

Foreword

Vulnerability and poverty are real threats to worldwide prosperity, and there is a need to act now. Fortunately this is exactly what this book does: presenting results from real life cases and simultaneously providing a methodology that can help us move forward towards increased awareness, improved understanding of risks, and effective risk management based on well-informed decision-making.

And this is why I was both honoured and pleased to accept the opportunity to write this foreword. During the past two decades I have been 'fighting' for a more sustainable development, focusing on how the combination of process management and decision-support can help stakeholders creating a better future together. My work has been both in developed and developing countries, and in all cases I have seen one key success factor: human beings. They create most of the problems, and they are the ones that have the ability to solve them too.

This brings me to my visit to Buenos Aires last year where not only the beautiful parks and the nicely revitalized harbour area were visited, but also 'La Villa de Retiro': a massive semi-illegal settlement of marginal dwellers. The contrasts are so striking between rich and poor, opportunities and threats, and this is not only the case in Buenos Aires, but in all Latin America.

After having seen the problems with my own eyes, we (the authors and I) continued the discussion on how a better understanding of real life problems can contribute to their solution, and I was again impressed by their knowledge but also in particular by their desire to make a real difference. They not only convinced me about the importance and urgency of the problems, but they also gave me hope for the future.

This hope is embedded in the contents of this book, because both challenges and a workable strategy towards solutions are presented in a coherent way. The particularly useful perspective, and emphasis of this book, is on the risks related to how social vulnerability and environmental hazards negatively reinforce each other in so-called 'hot-spots'. The methodology for identifying these risks even includes an attempt to take into account cumulative effects of localized environmental hazards e.g. caused by several small firms. In that respect the authors make a daring, and necessary, step towards improved inter-disciplinary approaches for addressing real life challenges. They acknowledge that the richness and complexity of life is much more than its individual parts, and, in particular, they focus on how understanding of the problems provides a basis for effective governability and management of risks.

Policy makers should therefore read this book focusing on how governability affects performance, and how the understanding of 'hot-spots' and GIS can be used together as a powerful tool for visualizing results and facilitating decision-making. Scientists should read it, because the methodology suggested represents a daring attempt to address real life cumulative problems and there is a need for more scientists to engage in inter-disciplinary science.

Finally, I would like to return to the three issues mentioned initially, increased awareness, improved understanding of risks, and effective risk management based on well-informed decision-making. These are interdependent objectives. Effective risk management builds on understanding, and awareness is an important step towards support for changes to policy and practice. This is why this book is so important, because it offers the basis for an integral approach to all three issues!

As the Latin American case studies especially illustrate, (much) more understanding of 'hot-spots' is definitely needed, but it should be accompanied by, and embedded in, an approach that focuses on multi-actor involvement. My hope and strong recommendation for the future is, therefore, that the stakeholders seize this opportunity: together.

Deventer, The Netherlands   Kjell-Erik Bugge

Acknowledgments

First of all we wish to thanks Spanish Agency for International Cooperation and Development (AECID). Without the grant this research would have not been possible. We are also immensely indebted to two people for their constant support and trust. Professor Jose J. Céspedes-Lorente has always been there providing encouragement and helpful suggestions while Professor Ken Peattie has given insightful comments, guidance and advice from the early stages of this research. Again, without their faith on the project, this book would have not been possible.

Special Thanks as well to Catherine Liston-Heyes and Kjell-Erik Bugge which made many useful suggestions and gave us valuable feedback and empathy. We would also like to thanks Rodrigo Lozano for his insights and critical analysis, Alejandro Martucci for his efforts to map industrial risks in Venezuela and in particular Clovis Zapata for his ideas for a spin-off of this project focusing on rural hot-spots in Brazil. We hope we will continue exploring joint avenues for collaboration in the future.

Finally, many thanks to Miguel Angel Plaza-Ubeda for his invaluable assistance in a variety of technical emergencies.

Contents

1 **Introduction** 1
José A. Plaza-Úbeda, Claudia E. Natenzon, Diego A.
Vazquez-Brust, Jerónimo de Burgos-Jiménez, and Julieta Barrenechea

2 **Evaluating the Firm's Environmental Risk:
A Conceptual Framework** 15
Diego A. Vazquez-Brust, Claudia E. Natenzon, Jerónimo de
Burgos-Jiménez, José A. Plaza-Úbeda, and Sergio D. López

3 **Statistical Information for the Analysis of Social
Vulnerability in Latin America – Comparison with Spain** 35
Anabel Calvo, Mariana L. Caspani, Julieta Barrenechea, and
Claudia E. Natenzon

4 **Evaluating the Firm's Environmental Hazardousness: Methodology** 53
Sergio D. López and Diego A. Vazquez-Brust

5 **The Case of Bolivia** 69
Luis Augusto Ballivián-Céspedes, Yolanda Bueno-Cachadiña,
and Sergio D. López

6 **The Case of Argentina** 91
Claudia E. Natenzon, Diego A. Vazquez-Brust, and Sergio D. López

7 **The Case of Spain** 117
José A. Plaza-Úbeda, Julieta Barrenechea, Jerónimo de
Burgos-Jiménez, Miguel Pérez-Valls, and Sergio D. López

8 **Concluding Remarks** 137
Jerónimo de Burgos-Jiménez, Diego A. Vazquez-Brust, José A.
Plaza-Úbeda, and Claudia E. Natenzon

Index 145

Contributors

Luis Augusto Ballivián-Céspedes Colegio Pestalozzi, Padilla 174, Sucre, Bolivia, auballi@entelnet.bo

Julieta Barrenechea Cátedra Sánchez-Mazas UPV/EHU, Universidad del País Vasco, 72 (20018) San Sebastian, Spain, ylabafej@ehu.es

Yolanda Bueno-Cachadiña OB Mallas SRL, Av. Jaime Mendoza 1202 A, Sucre, Bolivia, yolandabuenoc@yahoo.es

Jerónimo de Burgos-Jiménez Department of Business Administration, University of Almeria, Almeria 04120, Spain, jburgos@ual.es

Anabel Calvo Faculty of Philosophy and Letters, Institute of Geography "Romualdo Ardissone", University of Buenos Aires, Buenos Aires 1406, Argentina, belcalvodiaz@gmail.com

Mariana L. Caspani Department of Geography, Faculty of Philosophy and Letters, University of Buenos Aires, Buenos Aires 1406, Argentina, mcaspani@gmail.com

Sergio D. López Unidad de Coordinación de Programas y Proyectos con Financiamiento Externo: Programa de Infraestructura Hídrica de las Provincias del Norte Grande, Ministerio de Planificación Federal, Inversión Pública y Servicios, Av. Roque Saenz Peña 938, Piso 6, CABA CP 1035, sergiodlopez@yahoo.com

Claudia E. Natenzon Faculty of Philosophy and Letters, Institute of Geography "Romualdo Ardissone", University of Buenos Aires, Buenos Aires 1406, Argentina, natenzon@filo.uba.ar

Miguel Pérez-Valls Department of Business Administration, University of Almeria, Almeria 04120, Spain, mivalls@ual.es

José A. Plaza-Úbeda Department of Business Administration, University of Almeria, Almeria 04120, Spain, japlaza@ual.es

Diego A. Vazquez-Brust The Centre for Business Relationships, Accountability, Sustainability and Society (BRASS), Cardiff University, Cardiff Wales CF10 3AT, UK, VazquezD@cardiff.ac.uk

List of Acronyms

3D	Tridimensional
AECID	Spanish Agency of International Cooperation for Development
	Agencia Española de Cooperación Internacional para el Desarrollo
AL	Latin America
	América Latina
BRASS	The Economic and Social Sciences Research Council Centre for Business Relationships Accountability and Society at Cardiff University
CABA	Autonomous City of Buenos Aires
	Ciudad Autónoma de Buenos Aires
CBI	Confederation of British Industry
CCAA	Autonomous communities of Spain
	Comunidades Autónomas de España
EU (*CE*)	European Commission
CLACSO	Latinoamerican Council of Social Sciences
	Consejo Latinoamericano de Ciencias Sociales
CMDSN	National Social Minimum Data Sets
	Conjunto mínimo de datos sociales nacionales
CSR (*RSE*)	Corporate Social Responsibility
DAC (*CAD*)	Development Assistance Committee, OECD
EAGGF (*FEOGA*)	European Agricultural Guidance and Guarantee Fund
ECHP (*PHOGUE*)	European Community Household Panel
ECLAC (*CEPAL*)	Economic Commission for Latin America and the Caribbean
EPA	Environmental Protection Agency USA
ERDF (*FEDER*)	European Regional Development Fund
ESCB (*SEBC*)	European System of Central Banks
ESRC	The Economic and Social Sciences Research Council UK
ESS (*SEE*)	European Statistical System
EU (*UE*)	European Union
Eurostat	Statistical Office of the European Communities

FLACSO	Latin American Faculty of Social Sciences – Buenos Aires
	Facultad Latinoamericana de Ciencias Sociales
FTAA (*ALCA*)	Free Trade Area of the Americas
IDB (*BID*)	Inter American Development Bank
FIFG (*IFOP*)	Financial Instrument for Fishing Guidance
GIS (SIG)	Geographic Information System
IHD (*IDH*)	Human Development Index
IMF (*FMI*)	International Monetary Fund
INDEC	Statistical office of Argentina
	Instituto Nacional de Estadística y Censos
INE	Acronym of the Statistical Office of Bolivia and also of Spain
	Instituto Nacional de Estadística
IP	Index of Industrial Hazardousness (perilousness)
IR	Index of Evaluated Risk
ISDR	International Strategy for Disaster Reduction
ISO	International Organization for Standardization
ISV	Index of Social Vulnerability
LCS	Living Conditions Survey
	Encuesta de Condiciones de Vida
LEC (*NCA*)	Level of Environmental Complexity
	Nivel de Complejidad Ambiental
m.a.s.l	meters above sea level
MABA (AMBA)	Metropolitan Area of Buenos Aires
	Area Metropolitana del Gran Buenos Aires
NGO (*ONG*)	Non Governmental Organization
OAS (*OEA*)	Organization of American States
OECD (*OCDE*)	Organization for Economic Co-operation and Development
PAHO (*OPS*)	Pan American Health Organization, UN
PIRNA	Natural Resources and Environment Research Programme of Argentina
	Programa de Investigaciones en Recursos Naturales y Ambiente
PL (*LP*)	Poverty Line
RAMINP	Regulation on disturbing, unhealthy, toxic and dangerous activities
	Reglamento de actividades molestas, insalubres, nocivas y peligrosas
SABI	Database of Spanish and Portuguese Companies
	Base de datos de Empresas Españolas y Portuguesas
SEN	National Statistical System
	Sistema Estadístico Nacional
SIGBI	Soroptimist International Great Britain and Ireland
SME (*PyME*)	Small and Medium Entreprises
UBA	University of Buenos Aires

UBN (*NBI*)	Unsatisfied Basic Needs
	Necesidades Basicas Insatisfechas
UDAPE	Social and Economic Policy Analysis Unit Bolivia
	Unidad de Análisis de Políticas Sociales y Económicas
UN (*ONU*)	United Nations
UNDP (*PNUD*)	United Nations Development Programme
UNFPA	United Nations Population Fund
US SIC	United States Standard Industrial Classification System
USA	United States of America
WB (*BM*)	World Bank
WHO (*OMS*)	Organización Mundial de la Salud World Health Organization
WTO (*OMC*)	World Trade Organization

List of Figures

Fig. 1.1	Territory size by Iberoamerican countries (km^2)	7
Fig. 1.2	Population by Iberoamerican countries (millions of inhabitants)	8
Fig. 1.3	GDP per inhabitant (dollars)	8
Fig. 1.4	GDP by Iberoamerican country (million dollars)	9
Fig. 2.1	Conceptual model: dimensions of risk	17
Fig. 4.1	Evaluated hazard (EH) as a function of the distance from the industrial site	58
Fig. 4.2	Radius of influence of hazard (R) as a function of maximum EH	58
Fig. 4.3	Cumulative hazards at a given point calculated by totalling the impact of each industry as a function of its distance from that point	59
Fig. 4.4	3-D image of the bell of the kernel density function	60
Fig. 4.5	Comparison of the values given by the kernel and linear functions	61
Fig. 4.6	Kernel function applied to a single point and the sum of the effects of several points	62
Fig. 4.7	The effect of radius R on the surfaces, given the same initial points	62
Fig. 4.8	Density function based on the distance from the point and the category, adjusted in order that $q = 1$ at $R = 0$	64
Fig. 5.1	Map of industrial hazardousness for the departmental capitals of Bolivia	77
Fig. 5.2	Index of industrial hazardousness (IP) for the departmental capitals of Bolivia	78
Fig. 5.3	Map of social vulnerability for the departmental capitals of Bolivia	78
Fig. 5.4	Index of social vulnerability (IVS) for the departmental capitals of Bolivia	79
Fig. 5.5	Map of combined risk for the departmental capitals of Bolivia	80

Fig. 5.6	Index of combined risk for the departmental capitals of Bolivia	80
Fig. 5.7	Map of industrial hazardousness for Santa Cruz de La Sierra	82
Fig. 5.8	Map of industrial hazardousness for Sucre	83
Fig. 5.9	Map of social vulnerability for Sucre	85
Fig. 5.10	Map of social vulnerability for Santa Cruz de la Sierra	86
Fig. 5.11	Map of combined risk for Sucre	87
Fig. 5.12	Map of combined risk for Santa Cruz de la Sierra	88
Fig. 6.1	Map of industrial hazardousness per partido/department in Argentina	102
Fig. 6.2	Map of social vulnerability in Argentina by department/partido	105
Fig. 6.3	Map of evaluated risk in Argentina by departments/partidos	107
Fig. 6.4	Map of location of industrial activities that generate hazard	110
Fig. 6.5	Map of industrial hazardousness in the metropolitan area of Buenos Aires	111
Fig. 6.6	Map of social vulnerability in the metropolitan area of Buenos Aires by census unit (block groups)	112
Fig. 6.7	Map of evaluated risk in the metropolitan area of Buenos Aires by census unit (block group)	114
Fig. 7.1	Map of environmental hazardousness in Spain by municipalities	121
Fig. 7.2	Index of social vulnerability in Spain by municipalities	125
Fig. 7.3	Map of evaluated risk in Spain by municipalities	128
Fig. 7.4a	Map of social vulnerability in Madrid by census unit	130
Fig. 7.4b	Map of social vulnerability in Seville by census unit	130
Fig. 7.5a	Map of industrial hazardousness in Madrid	131
Fig. 7.5b	Map of industrial hazardousness in Seville	131
Fig. 7.6a	Map of evaluated risk in Madrid by census unit	132
Fig. 7.6b	Maps of evaluated risk (IR) in Seville by census unit	132

List of Tables

Table 1.1	Cross-country data	7
Table 2.1	Evaluated industrial risk: the combination of hazardousness index with vulnerability index	22
Table 2.2	Ranges of evaluated industrial risk	23
Table 2.3	Poverty and environmental deterioration in Latin-America	24
Table 3.1	List of selected indicators	41
Table 3.2	Results of the compilation and selection of indicators for Latin America and the Caribbean	44
Table 3.3	Dimensions, variables and indicators for Argentina and Spain	51
Table 4.1	Values of R adopted depending on the category	63
Table 4.2	Weighting factors	63
Table 5.1	General data on the departments of Bolivia	70
Table 5.2	Number of firms per department	76
Table 5.3	Environmental Hazard Contribution by industrial sector in Sucre	81
Table 5.4	Environmental Hazardousness Contribution by industrial sector in Santa Cruz de la Sierra	81
Table 6.1	Annual emission factors per industrial sector (tonnes/employee)	97
Table 6.2	Index of industrial hazardousness (IP): ranges and frequencies	97
Table 6.3	Radius of influence of the hazardousness of an industry	99
Table 6.4	Social vulnerability in Argentina: indicators used	100
Table 6.5	Index of Social Vulnerability (ISV): ranges and frequencies	100
Table 6.6	Industrial emissions of contaminant particles in Argentina	101
Table 6.7	Departments with high and very high Index of Industrial Hazardousness (IH)	104
Table 6.8	Departments with higher Index of social vulnerability (ISV)	106

Table 7.1	Municipalities and resident population according to levels of IP	122
Table 7.2	Spanish municipalities with high or very high IH and the population exposed	122
Table 7.3	Dimensions and indicators of social vulnerability in Spain	123
Table 7.4	Spanish municipalities with very high ISV	125
Table 7.5	Municipalities and resident population according to levels of ISV	127
Table 7.6	Municipalities and resident population according to levels of IR	128
Table 7.7	Descriptive data of the municipalities analysed on the scale of census units	129

About the Contributing Editors

Jerónimo de Burgos-Jiménez, is a lecturer in Business Organization at the Universidad de Almería. He specializes in Operations Management, environmental management and stakeholders. He has published several works that focus on firms' environmental management and production management in different international journals such as Omega, Scandinavian Journal of Management, International Journal of Operation and Production Management, Services Industries Journal, Journal of Business Ethics or Business Strategy and the Environment. In the research line of the present work he has written chapters for other international books, as well as contributing to congresses and publishing articles in related scientific journals.

Claudia E. Natenzon Geographer, graduated with honours in 1975 from the Faculty of Philosophy and Artsat the Universidad de Buenos Aires. Doctor cum laude in Geography from the Universidad de Sevilla, España (2000). She is a lecturer at the Faculty of Geography and Arts, UBA. As a researcher at the Geography Institute of that Faculty since 1988 she has directed the Research Program on Natural Resources and the Environment Recursos Naturales y Ambiente-PIRNA. She has specialized in issues of environmental risk, particularly the diagnosis of social vulnerability and its relationship with catastrophes. In recent years she has applied this knowledge to projects linked to social vulnerability related to climate dynamics, with national (UBA, Agencia, CONICET) and international financing (OEA, PNUD, PNUMA, GEF, NSF). In the Argentinian Headquarters of FLACSO – Latin American Faculty of Social Sciences – she has participated in the Planning and Associated Management Program as an associate researcher and postgraduate lecturer since 1988. She has worked for national public organisms (DNOA, APN, CFI, CONICET, Civil Defence, DNCC) and international ones (FAO, Politécnico de Milán, ALADI, BM, PNUD, GEF).

José A. Plaza-Úbeda, is a lecturer and Doctor in the Area of Business Organization at the Universidad de Almería, specializing in stakeholder management and firm environmental management. He presented his thesis on stakeholder integration in firms in 2005 and had previously published articles related to stakeholder management in international journals such as Journal of Business Ethics and Business

Strategy and the Environment. He has been the Spanish coordinator of the project financed by the Agencia Española de Cooperación y Desarrollo that has given rise to the present work. In the line of research of the present work, he has previously contributed chapters in international books, as well as participating in congresses and publishing articles in related scientific journals.

Diego A. Vazquez-Brust Civil Engineer, MBA and Doctor in Economics and Business from Royal Holloway, University of London (United Kingdom). Administration manager of environmental projects in the province of Buenos Aires (1997–2002); consultant on issues relating the firm and the environment for international public organizations (IDB, PNUD, WAG) and private ones (YKK.SA, Kirin SA.). Since 2007 he has been Research Manager at BRASS (Cardiff, UK) where he coordinates research projects on Poverty, Vulnerability and Industrial Risk; and Corporate social responsibility in the mining industry. He has lectured Global Economy at Royal Holloway, University of London since 2006. Since 2009 he is one of the global coordinators of GIN (Greening of Industry Network) – an international network which promotes the integration of environmental issues in firms, government and academia. His recent work include the book 'Collaboration Facilitating Sustainable Innovation through Collaboration' (Springer, 2010) with Sarkis & Cordeiro and articles on environmental risk and social responsibility in Latin America (Journal of Environmental Management and Journal of Business Ethics).

About the Contributors

Luis Augusto Ballivián-Céspedes Chemical Engineer, Universidad Mayor, Real y Pontificia de San Francisco Xavier de Chuquisaca, Bolivia. Specialist in the Educational Direction of Innovation, UNESCO – Universidad Privada del Valle, Bolivia. Specialist in Management of Production, Quality and Technology, Universidad Politécnica de Madrid, Spain. Master in Higher Education – Universidad Andina Simón Bolívar, Bolivia. Master in Communication and Educational Technologies – Instituto Latinoamericano de la Comunicación Educativa, México – Universidad Andina Simón Bolívar, Bolivia. Undergraduate and postgraduate lecturer at Universidad del Valle and Universidad Andina Simón Bolívar. President of the Educational Society Juan Enrique Pestalozzi.

Julieta Barrenechea Sociologist (UBA, 1994). Diploma of Advanced Studies in Science and Technology Management from the Universidad del País Vasco (UPV/EHU), 2007. Associate researcher with PIRNA – Natural Resources and Environment Research Program – (UBA) and member of the Area of Research and Management of Science, Technology and Innovation Networks Cátedra Sánchez-Mazas UPV/EHU. She has been a research grant fellow at CONICET (National Council for Scientific and Technical Research, Argentina) and technical advisor to the Presidency of the Commission for Science and Technology of the Chamber of Representatives of the Nation of Argentina. She has conducted research into management and social communication of technological risks, methodologies of measurement and analysis of social vulnerability to risk in nationally funded projects (CONICET, UBACYT, ANPCyT) and international ones (GEF, CIDA). At present she is carrying out research and projects of public policies linked to new patterns of scientific activity, assessment and management networks of knowledge based on the quality of relations, the development of competences for collaborative work and governance financed by the Basque Government, Diputación Foral de Gipuzkoa and FECyT Spain.

Yolanda Bueno-Cachadiña Yolanda earned a degree in Industrial Engineering with specialization in Electronics from the University of Extremadura. She further obtained a degree in Management from the same university before moving to Bolivia where she completed postgraduate studies in Business Administration at Universidad Andina Simon Bolivar (Sucre,). She is now a doctoral researcher at Universidad de Almería (Spain) and Universidad Andina Simón Bolivar where she participates in a research programme investigating industrial clusters, project development and environmental management. Yolanda is an experienced project consultant and has been involved with a range of public and private developmental projects in Bolivia.

Anabel Calvo Geographer, graduated in 1987 from the Faculty of Philosophy and Arts of the Universidad de Buenos Aires. She is currently finishing her Master in Economic Sociology at IDAES, Universidad Nacional de San Martín. She is a lecturer in the Faculty of Philosophy and Arts, UBA, and is currently working as JTP in Latin American Social Geography and Resources and Society. She is a teaching mentor in the School of Training and Skills (CEPA) of the government of the City of Buenos Aires. As a researcher in training she is part of the Research Program on Natural Resources and Environment, PIRNA, of the Geography Institute of the UBA. She has specialized in issues of environmental risk, in projects financed by the UBA particularly on topics related to risk management.

Mariana L. Caspani Geographer (UBA, 1989). A lecturer in the above-mentioned faculty, as a JTP in Latin American Social Geography. She is a tutor of foreign students in the CIEE exchange programme, in the Argentinian headquarters of FLACSO (Latin American Faculty of Social Sciences). She is responsible for GIS in the General Electoral Committee of the Government of the Autonomous City of Buenos Aires-GCABA. Independent consultant on Geographical Information Systems. She has specialized in applying GIS to different problems, particularly regarding issues of urban planning, electoral processes and results, management of commercial territories and indicators of socioeconomic level and living standards. In this line she has worked with state organisms (UBA, GCABA, BID, PNUD) and as a member of private companies linked to projects of digital cartographic development and consulting on spatial analysis. She has participated in projects of the Geography Institute at the Faculty of Philosophy and Arts in the Universidad de Buenos Aires and has given classes in the Postgraduate Programme of Urban and Regional Planning of the Faculty of Architecture and Town Planning at the UBA in the School of Training and Skills (CEPA) of the GCABA.

Sergio D. López has degrees in Construction Engineering (1993) and Civil Engineering (1995) at the Universidad Nacional de La Plata; and a Master in Environmental Management and Auditing, at Universidad de Palma de Gran Canaria (2004). A specialist in GIS and Remote Sensors, he has worked as a consultant in several projects in both the public and private sectors. In the public sector: since 2008: he is specialist in hydraulics in the Hydric Infrastructure Program in Norte Grande, Ministerio de Planificación Federal. Before, he worked for the Environmental Policy Agency of Buenos Aires Province as Director of the Program 'Environmental Information and Planning System' (2004–2005). In the private sector he is currently Director of Engineering in the company Aetos Consultores. He has participated in several projects of engineering, environment and information systems. During his stay in the Environmental Policy Agency, he designed a system of spatial indicators of hazardousness using GIS which he later applied in Wales in 2007 as a visiting BRASS scholar at the University of Cardiff.

Miguel Pérez-Valls Doctor from the Universidad de Almería and a graduate in Economics and Business Sciences, he is currently a collaborating Doctor and lecturer in Business Organization at the Universidad de Almería. His doctoral thesis from 2009 analyzes flexible organizational forms and the development of organizational capabilities. He collaborates actively with different non-governmental organizations to promote international development and has wide experience of running/managing such organizations. In the line of research of the present work, he has contributed chapters to international books, participated in related scientific congresses and participated actively in analyzing the Spanish case in the framework of the AECI project which forms the basis of this work.

Chapter 1
Introduction

José A. Plaza-Úbeda, Claudia E. Natenzon, Diego A. Vazquez-Brust, Jerónimo de Burgos-Jiménez, and Julieta Barrenechea

Abstract This chapter introduces the research project whose results are summarised in the book. It describes how poverty and environmental degradation influence each other in areas where vulnerable populations are exposed to environmental hazardousness generated by industrial activities ('hot-spots'). It outlines the long-standing conceptual gap in research when addressing the 'vicious circles' between poverty and environmental deterioration – a major challenge to sustainable development for business and societies. Seeking to address such challenge, the project is anchored within the perspective of sustainability science, the emerging field of user-inspired research exploring the interactions between human and environmental systems. The chapter then presents the geographical area of study: Ibero-America, providing a description of historical, cultural, and economic Spain-Latin-America relationships. Finally the chapter provides summaries and linkages amongst the remaining chapters.

Keywords Sustainability science · Poverty · Environmental degradation · Vicious circles · Ibero-America

1 Institutional Background

The present work is based on the research project entitled "Firms' Environmental Impact, Social Vulnerability and Poverty in Ibero-América: Analysis of Interaction and Diagnosis of Areas of Potential Risk" led by the University of Almería[1] (Spain), The University of Buenos Aires (Argentina) and the BRASS[2] research centre of Cardiff University (United Kingdom). The project has been carried out by a

---

[1] This work has been partially funded by the Spanish Ministry of Science and Education and the European Fund for Regional Development (reference ECO2008-03445/ECO).

[2] BRASS (Business Relationships Accountability, Sustainability and Society) at Cardiff University is a major UK Economic and Social Research Council funded (ESRC) Centre that was launched in 2001 as a combination of Cardiff Business School, Cardiff Law School and the School of City and Regional Planning.

J.A. Plaza-Úbeda (✉)
Department of Business Administration, University of Almeria,
Almeria 04120, Spain
e-mail: japlaza@ual.es

joint team of researchers in Spain, Argentina, Bolivia, Venezuela, Brazil and the United Kingdom over 2008 and 2009. Travelling expenses of the Latin American researchers, acquisition of the databases for elaborating the maps presented and the publication of the results were funded by a grant from the Spanish Agency for International Cooperation awarded to the University of Almería in coordination with the University of Buenos Aires (References A/9527/07 and A/017735/08). The project also received funding from the Socrates-Erasmus Programme to facilitate the mobility of researchers between the Universities of Almería and Cardiff, plus the institutional support of all the universities involved: Universidad de Almeria, Universidad de Buenos Aires, Cardiff University, Universidad Simon Bolívar (Sucre), Universidad de Caracas and Universidad de Brasilia.

This research project was made possible thanks to the coordination between the researchers in each of the above-mentioned countries. The research teams' interdisciplinary groups consisted of researchers and practitioners from different branches of knowledge: Geography, Sociology, Business Studies, Economics, Engineering, Information Technology, Environmental Sciences, Biology and the Humanities.

The diversity of the zones of study has also meant that different results were to be obtained in each of the countries studied, conditioned in the main by access to information and limitations on resources. The different research teams in each country have carried out their work independently, though meetings were held to coordinate efforts in Spain and Argentina in 2008 and 2009, respectively. These meetings helped the teams to agree on many issues: the methodological approach, the terminology used, the extent of the work in each country, solutions for the challenges that arose, and ultimately the aims and structure of this book.

2 Conceptual Background: Sustainability Science and the 'Vicious Circle Poverty-Environmental Deterioration'

Sustainability Science is as yet a developing field (Kates and Dasgupta 2007). It can be described as a discipline that produces knowledge on the complex interactions between natural and social systems and their roles in affecting the planet's sustainability (Kua and Ashford 2004). As such, Sustainability Science aims to develop practical solutions to real sustainability challenges though a new research paradigm that breaks down artificial divides between the natural and social sciences, and between knowledge generation and its practical application in decision-making.

Drawing on systems dynamics, sustainability science literature warns against policy or research downplaying interactions between economic and social systems (Kates et al. 2001; Palmer et al. 2005) The transition to sustainability lies precisely in the acknowledgement of the intertwined nature of environmental issues and human activities (Clark and Dickson 2003). Their previous conceptualisation as largely separate and distinct is misguiding and has further obscured the fact that uncontrolled economic growth is a major menace to ecological and social systems.

However, the notion that environmental problems and developmental problems can, and indeed should, be tackled separately is deeply ingrained in a still dominant mindset that invokes the existence of trade-offs between environmental and social issues. One of the most cited development theories, known as Kuznet's curve, argues that the most effective policy to fight poverty and decrease environmental deterioration is to focus solely on the promotion of continued economic growth based on market mechanisms. Although both distributive inequality and environmental deterioration increase during the initial stages of rising economic development due to changes in technology and economic structure, once a certain degree of development has been reached inequality and the environmental impact start to diminish.

The practical implication of Kuznet theory is the promotion of uncontrolled economic growth. However, the validity of the theory in current conditions of environmental deterioration and economic globalisation is strongly questioned by Sustainability Science. The theory worked while the post World War 2 economical and technological paradigm was dominant. Nowadays, changes in technology, expanding the current structure of production and consumption will further aggravate environmental deterioration and pose a serious challenge to ecological integrity and social cohesion. The effects are accumulative and in many cases, as in the extinction of species, irreversible (Kates et al. 2001).

Therefore, it is becoming widely accepted that environmental deterioration, poverty and social inequalities are interlinked and that poverty reduction ought to be addressed in conjunction with environmental preservation and distributive justice (Sachs 2004). However, success with integrated strategies has been elusive (Kates and Dasgupta 2007). The intertwined nature of poverty and environmental deterioration is often described as a 'Vicious Circle' (Taylor 2008). Poverty brings about environmental decline (Hart 1995), which in turn increases the poverty of populations in vulnerable ecosystems or in those that suffer high levels of contamination due to human activity (Gray and Moseley 2005), where the productivity of soil use decreases or the cost of protecting human health increases.

Poverty increases environmental deterioration in both rural and urban areas. In the former, intensive agriculture, the use of fertilisers or the felling of forests lead to deforestation, erosion of the topsoil and the contamination of water sources, all of which are exacerbated by the incapacity to invest in the environment and demographic pressure due to the parallel decline in purchasing power and in the birth rate (Hart 1995). In poor urban areas environmental regulation tends to be weaker (Pargal and Wheeler 1996), partly since the poor are less well-informed of the risks, partly because they are less able to apply pressure to improve environmental quality, and partly because they place more relative importance on the possibility of employment than on protecting the environment (Dasgupta et al. 1998). This leads to a greater density of 'dirty', inefficient and contaminant industries (Hettige et al. 1998).

This trap locks populations in developing regions into a situation with a narrower margin for survival, increased vulnerability to natural hazards, and increasing fragility of the ecosystems on which the residents depend (Adger 2006). All of these factors are exacerbated by a lack of capital or technological investment, a lack of work skills among residents, inadequate education, and poor governance (Taylor 2008). The vulnerability of these populations may be further exacerbated by unjust or ineffective policies (Swart et al. 2004).

A second aspect of the vicious circle between poverty and deterioration refers to relations between increases in wealth and environmental deterioration. Poverty reduction achieved at the expense of environmental deterioration is unsustainable and leads to long term increases in inequality (Hart 1995). On the one hand, when economic growth is achieved at the expense of flexibility in the enforcement of environmental regulations, in 'dirty' sectors there is increased risk of environmental contamination and industrial accidents that threaten the health, quality of life and economic standing of the surrounding population. Furthermore, there is actually an increase in the effective poverty of those who have little recourse to protect themselves from environmental diseases, who find that their health can be affected and they can lose the opportunity to work and contribute to sustaining the family nucleus (Gray and Moseley 2005).

Arguably, an increase in wealth will eventually lead to more environmentally aware populations (such as those in the west), which in turn will put pressure on policy-makers to improve environmental control. However, lack of governance and insufficient access to information about risks in the more vulnerable populations may raise the bar in terms of the economic growth required to trigger such 'awakening'. Reaching such levels of wealth all over the world would cause irreversible global environmental deterioration due to the increase in emissions and the greater use of natural resources linked to the increase in purchasing power. If all developing countries were to reach western levels of consumption, four earths would be needed to support their needs (Baker 2006).

A person's ability to contribute to solving such challenges depends on whether the individual has the opportunity and the willingness to make the behavioural or ideological changes needed to make the contribution successful (Kua and Ashford 2004). Willingness to change is usually preceded by reflection on the impacts our actions have on nature and on society as a whole. In turn, the preamble for such reflection is awareness of the connections between behaviour that is taken for granted (e.g. environmental enforcement officers in Argentina focus on big polluters and turn a blind eye to pollution from small firms since they see it as 'insignificant'(Vazquez-Brust and Liston-Heyes 2010) and the threat that such behaviour represents to particular places and social groups. Here, we present the results of a study that illustrates how environmental and economic perspectives can be integrated in a practical approach to diagnose areas in need of intervention. This approach brings visibility to affected populations and can therefore set awareness in motion, in turn triggering policy and community actions to improve both nature and society, thereby eradicating the 'poverty trap'.

The main aim of the present work is to contribute knowledge to an on-going line of research which intends to analyse the potential risk in a given geographic area as a result of socio-demographic characteristics and of industrial activity. Throughout the work this risk is referred to as 'evaluated risk' or 'potential risk', and as such it can be used in a homogeneous methodology applied to different zones of analysis which allows the identification of areas of high or very high levels of risk. Once these areas are identified, preventive measures can be devised to avoid possible future environmental or social catastrophes. In order to operate within this context a specific methodology is developed that can be applied in different geographic areas, and which is intended to analyse and identify the areas with the greatest problems of social vulnerability and environmental hazardousness.

These areas are referred to as 'hot-spots' throughout the work. Intervention models based on identifying specific areas of risk or hot-spots constitute a specific work methodology directed towards analysing potential risk, and they have two main advantages. Firstly, they allow us to compare different areas of study thanks to the homogeneous indicators used, and secondly, they allow us to analyse the situation in greater depth. The present work follows both these lines for each case studied, carrying out an analysis to locate those municipalities at greatest potential risk in the whole country, but also applying the methodology on a lower scale, namely the census unit. This level of analysis allows us to identify more specific areas of study, for which the characterisation of sources of hazard and the design of preventive measures is, a priori, more efficient than other more general measures suggested on the basis of a more global analysis.

Sustainability science research is based on five pillars: (a) aiming to advance understanding of a 'grand challenge' or observed problem while at the same time providing practical policy tools; (b) effective solutions to observed problems should consider the economic, environmental and social factors that contribute to the problem; (c) problem identification and solution formulation should be place-based and span across all appropriate spatial and temporal scales; (d) applying an integrated approach comprising of qualitative and quantitative methodologies; (e) integrating the views from a wide range of scientific disciplines in an interdisciplinary and international approach (Kates et al. 2001; Swart et al. 2004).

This book is sustainability science, albeit at an early stage of development. It illustrates sustainability science because it is clearly interdisciplinary, with lead authors and contributors from economics, geography, management science, engineering and environmental sciences. It is international: the lead authors and editors all come from different countries and have all worked in developing countries. It is place-based and seeks to identify problems on a variety of geographical scales. It tackles a grand challenge rivalled in our time perhaps only by climate change, peace and security (Kates and Dasgupta 2007). Most importantly, it is an example of sustainability science because it asks fundamental questions but seeks practical and place-based solutions. Finally, to further the success of such policy solutions, problem identification has been carried out in collaboration with local researchers and stakeholders who are familiar with the problems in question.

3 The Geographical Area of Study: Ibero-America

The concept of Ibero-America refers to the countries that constitute the Organisation of Ibero-American States, namely Spain, Portugal, Andorra and 19 Latin American and Caribbean countries: Argentina, Bolivia, Brazil, Colombia, Costa Rica, Cuba, Chile, Ecuador, El Salvador, Guatemala, Honduras, Mexico, Nicaragua, Panama,

Paraguay, Perú, the Dominican Republic, Uruguay and Venezuela. The definition of Ibero-America arose from the Ibero-American Summits, which have brought together the chiefs of state and heads of government of the above-mentioned countries since 1991. The Declaration of Guadalajara of 1991, which was the result of the first summit, and the Declaration of Madrid of 1992 are considered its founding charters. This group of countries account for a quarter of the world population and 11% of global wealth (taken as a % of worldwide purchasing power). Of this 11%, 1.8% corresponds to Spain, 2.88% to Brazil, 2.1% to Mexico, 0.82% to Argentina and 0.5% to Bolivia (Malamud et al. 2011).

Ibero-America is considered an appropriate area of study since it is a heterogeneous group of countries with a common set of historical and cultural characteristics. Management practices and government policies have evolved in a similar fashion in many of these countries over recent years. The historical links between Spain and Latin America which date from over 500 years ago are still of great import today from a social, economic and political viewpoint. Indeed, the Spanish Constitution establishes in Article 11.3 that 'The state shall grant treaties of dual nationality with Latin American countries or with those countries that have had or still have a special link with Spain'. Although the importance of social and economic relations between Spain and Latin America has varied over the years, in recent times this link has strengthened considerable. By way of example we should mention the Ibero-American Summits held regularly between the 22 member states since 1991.

The following table and graphs illustrate the comparative situation of the 21 Ibero-American countries regarding different socioeconomic variables which put the current scenario into context. The three countries chosen for the empirical framework of this project (Argentina, Spain and Bolivia) are seen to present similar characteristics, but also differences. This makes them appropriate choices for the present study, which focuses on assessing major differences related to the risks of social vulnerability and environmental hazardousness as a result of commercial activity.

Table 1.1 shows the Human Development Index (HDI) data for each country, the population, surface area and gross domestic product at constant prices (GDP). The situation of the countries analysed in the study (Spain, Argentina and Bolivia) in relation to the remaining Ibero-American countries is presented in Figs. 1.1, 1.2, 1.3 and 1.4.

The HDI is an internationally accepted measure of countries' degree of development. The data in Table 1.1 indicate that Spain is the country with the highest degree, Argentina is among those with a high degree, and Bolivia is among the countries with a low degree of development.

The choice of large countries is an important factor given the framework of the present study, since it is more conducive to identifying different problematic areas. After Brazil, Argentina is the largest Ibero-American country. While Bolivia and Spain are smaller, they are quite sizeable in comparison to the remaining Ibero-American states (see Fig. 1.1).

1 Introduction

Table 1.1 Cross-country data

PAIS	IDH	IDH Level (among 169 countries)	Territory size (km^2)	Population (inhabitants)	GDP (millions dollars)	GDP per Inhabitants (dollars)
España	0.863	20	504,750	47,021,031	1,464,040	31,946
Portugal	0.795	40	91,905	10,637,713	233,478	21,970
Chile	0.783	45	2,006,096	17,094,275	161,261	9,516
Argentina	0.775	46	3,761,274	40,091,359	310,057	7,725
Cuba	0.863	51	110,922	11,242,628	62,278	5,559
Uruguay	0.765	52	176,215	3,358,584	31,511	9,420
Panama	0.755	54	74,979	3,405,813	24,859	7,175
Mexico	0.75	56	1,964,375	112,336,538	874,810	8,133
Costa Rica	0.725	62	51,100	4,563,538	29,318	6,345
Peru	0.723	63	1,285,216	29,462,000	126,766	4,356
Brazil	0.699	73	8,547,403	190,732,694	1,567,823	8,220
Venezuela	0.696	75	916,445	28,833,845	325,678	11,382
Ecuador	0.695	77	256,370	14,306,876	57,503	4,059
Colombia	0.689	79	1,141,748	45,508,205	232,403	5,167
Republica Dominicana	0.663	88	48,671	9,884,371	46,714	4,815
El Salvador	0.659	90	21,040	6,183,002	21,101	3,623
Bolivia	0.643	95	1,098,581	10,426,154	17,464	1,707
Paraguay	0.64	96	406,752	6,340,639	14,216	2,265
Honduras	0.604	106	112,492	7,876,662	14,268	1,910
Nicaragua	0.565	115	120,339	5,742,300	6,149	1,070
Guatemala	0.56	116	108,889	14,361,666	37,661	2,687

Source: Malamud et al. (2011)

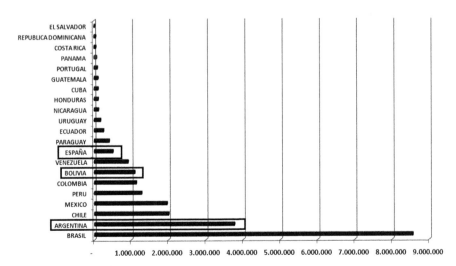

Fig. 1.1 Territory size by Iberoamerican countries (km^2)

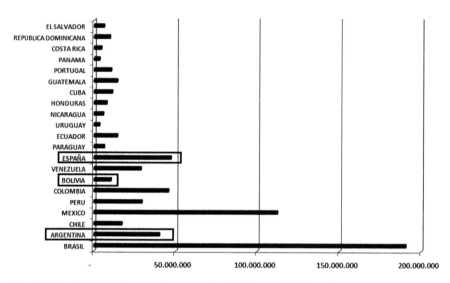

Fig. 1.2 Population by Iberoamerican countries (millions of inhabitants)

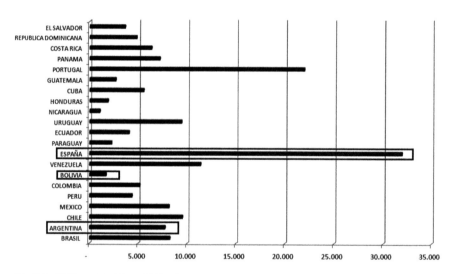

Fig. 1.3 GDP per inhabitant (dollars)

The dispersion or concentration of population can be relevant factors in the analysis of socioeconomic and environmental differences. As such, the choice of countries with differing population characteristics can also provide interesting results which allow us to identify determining factors. As can be seen in Fig. 1.2, Spain and Argentina have larger populations than Bolivia.

Figures 1.3 and 1.4 present the GDP of Ibero-American countries in absolute and relative terms. These are the most marked differences between the countries

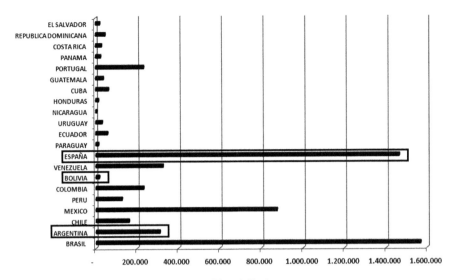

Fig. 1.4 GDP by Iberoamerican country (million dollars)

included in the study, and Spain has the highest GDP in both absolute and relative terms. Argentina is among the countries with higher GDP (especially in absolute terms), whereas Bolivia's GDP is much lower than that of the other two countries. These circumstances, together with the HDI data, have influenced the choice of these countries for analysis, since the level of economic development is one of the determining factors when assessing different risk levels of social vulnerability and industrial hazardousness.

It should be noted that historically relations between Spain and Latin America have had their ups and downs. Currently the situation is undergoing a major growth process. Spain devotes one third of its overseas investment to Latin America, and from 1996 it was the main European investor in this area (Platt 2000). In recent years Spanish investment has continued at high levels (Catan and Lyons 2008; Moffet and Prada 2010), and many Spanish multinationals are major players in Latin America (Repsol YPF, BBVA, Banco Santander, whose main assets are located in en Argentina, Venezuela, Bolivia and Ecuador. As Latin American nations opened up their economies in the 1990s, Spanish companies snapped up formerly government-owned concerns. From 1993 to 2008, Spanish firms had invested 130 billion euros ($165 billion) in the region – more than a tenth of Spain's annual GDP (Catan and Lyons 2008). Spanish companies are more heavily involved in sectors such as banking, financial services, telecommunications and energy (Jones 2001; Valdaliso 2008). These sectors have been liberalised in Latin America and the need for foreign investment has increased greatly. In this scenario of growing Spanish investment, there is also an increase in partnerships, joint-ventures and trade relations between Spanish/Portuguese and Latin American companies. In turn, the management style of the Spanish multinationals is more and more present in the social and economic life of Latin America. Indeed, previous studies have revealed great similarities in the ethical behaviour of management of the major Spanish and Latin American companies (Melé et al. 2006). Some of these common features include a growing trend for companies to implement formal ethics statements and the fact that large corporations are more likely to implement ethics statements than small and medium-sized companies.

The influence of other mechanisms of regional integration and cooperation (EU, MERCOSUR, etc.) notwithstanding, the definition of Ibero-America, and therefore membership of the OIS, is based on the concept of a common identity, defined in the main by the historical, cultural and particularly linguistic links. From the outset, the Ibero-American Summits have promoted agreement and cooperation inspired mainly in the conservation, consolidation and development of historical values and the cultural heritage of this group of countries, with explicit respect for the diversity which characterises them (Arenal Moya 2009). Over the years, cooperation promoted in the name of the Ibero-American community has become institutionalised, revolving around the central axis of 'development' to ensure suitable insertion of the member states on the international scene. The policies and initiatives promoted by the Ibero-American Summits are always in line with those of the United Nations.

In parallel with the institutionalisation process of Ibero-American cooperation, there has been consolidation of programmes which take on board challenges and problems that might require or be benefited by concerted actions. In this way, Ibero-American multilateral intervention has reached beyond its usual scope, consolidating its common historical values and cultural links.

This concept of development was developed in the first declaration of the Summit of Guadalajara (1991). It is associated with subjects such as multilateral commerce, narcotics dealing, women's issues, native population, disarmament, environment, health and agrarian development. In recent years we should add to these concepts social inclusion, innovation and knowledge (Quindimil 2010). It should also be mentioned that in parallel with the above-mentioned government action, a significant number of non-governmental associations and organisations also promote and celebrate forums covering a wide range of issues: parliament, local government, commerce, environment, nuclear issues, etc.

Regarding the subject matter of this book, it is interesting to contextualise the treatment of the environmental issue in the framework of Ibero-American cooperation policies. Bearing in mind the fact that the central theme is the concept of development, the environment has played a significant role since the Madrid Summit of 1992, which confirmed the commitment to 'sustainable development' proposed at the UN Conference on Environment and Development held in Río de Janeiro in 1992. Other important landmarks are the Ibero-American Congresses on Environmental Education which have been held since 1996, and in relation to environmental risk, the 2nd Congress held in 1997 specifically mentions the need to 'Inform, enable, orientate and make citizens aware, by means of specific programmes aimed at different sectors of society (industry, government, education, the media and society as a whole), of the cycles of nature and how they are manifested on a local scale and of the conditions of deterioration, in order to adapt technologies and to promote a culture of prevention which will help to understand and manage the risks that are faced', (Trelles et al. 1997). The Ibero-American Summits have currently adopted the UN Millennium Development Goals as a framework for their development strategy. Goal 7 of these refers to guaranteeing environmental sustainability. On the other hand, in 2001 the 1st Ibero-American Forum of Environment Ministers was held. In 2002 it was decided to give greater relevance to this topic, and since then this Forum has been held in parallel with the Summit of Chiefs of State.

Part of the content of the Declaration of Asunción (2008) is also relevant. In this declaration the Environment Ministers agree to 'reinstate environmental quality as one of the management priorities for governments in the region' and to 'emphasise the importance of environmental quality, including the suitable, integral management of substances, materials and residues, with the aim of improving the models of management and development, and the adoption and use of appropriate technologies for treating and/or recycling these residues, thus allowing further application of international agreements and commitments regarding chemical substances, and fomenting and reaching the necessary synergy between them, including the adoption of national regulations on these matters'.

In short, the Ibero-American context includes environmental questions among its areas of action for cooperation and development, and over the years more importance has been given to these issues, including commitments and initiatives in this field in the institutional framework. On the other hand, scientific and technological cooperation is also a key factor in the Ibero-American context. As such, the Declaration of the Ibero-American Summit of Salamanca (2005) explicitly states the 'commitment to advancing towards the creation of an Ibero-American Knowledge Space' whose aim is to consolidate Ibero-American interaction and collaboration between universities, research centres and companies with a view to generating, transmitting and transferring knowledge based on research, development and innovation.

In this scenario it would seem pertinent to carry out scientific research that allows information to be obtained and comparisons to be made between the countries and/or regions that make up this entity of international cooperation. The present study is therefore relevant for the subject matter itself, but also because its conclusions are the result of the collaboration between research teams from different universities, the majority of them in Ibero-American countries.

The book aims to elaborate and validate a methodology to assess risk in different Latin American countries and Spain by combining environmental, socio-economic and geographic concepts. To this end, spatial and technical indicators are devised to quantify the social vulnerability and environmental hazardousness of a given territory. By aggregating these indicators we have been able to quantify the evaluated risk in Argentina, Bolivia and Spain due to environmental deterioration. Our underlying hypothesis is that risk due to environmental deterioration in a given area must be analysed together with that area's vulnerability.

The work is divided into different chapters which deal with the aims put forward. Chapter 2 is organised into three sections. The first two sections outline the conceptual framework and empirical methodology used in this project to evaluate the environmental risk generated by the firm. Risk is defined as the result of combining potential hazard, vulnerability and exposure. The framework suggests that the gap separating real (or managed) risk and potential (or evaluated) risk widens with less uncertainty and greater governability. The third section provides a description of the governability of environmental impact in Latin America, where the divide between real and potential risk is low, and where, therefore, methodologies that evaluate potential risk may also be appropriate for interpreting the extent to which evaluated risk is being managed.

Chapter 3 shows the results obtained in the search for empirical information on social aspects in Latin America relating to social vulnerability on a national scale. It also puts forward the reasoning behind this methodological approach, the procedures employed in the search and the problems encountered, establishing where possible a comparison with the Spanish case. It shows the results obtained from the selection of indicators and the data corresponding to each of them for the vast majority of Latin American countries. It also compares the available indicators in Spain and Argentina covering the dimensions and the representative variables of different conditions of social vulnerability.

After a brief introduction summarising the dominant approach to development of risk maps and their relationship to the conceptual perspective used in this project, Chapter 4 details the empirical procedure used for calculating industrial hazardousness maps. This methodology measures the sum of potential hazard in a given geographical area, using an algorithm to extend the influence of the potential hazard of each industry to the surrounding area, also overlapping the effects of various industries within an area of influence. This indicates the location of areas of potential hazardousness due to the cumulative effects of small and medium-sized firms in each area that had not been identified by previous methodologies based only on the size or potential impact of individual companies.

Chapter 5 evaluates risk, social vulnerability and industrial hazardousness in Bolivia applying the methodology described in Chapters 3 and 4. As well as presenting aggregated risk results at the departmental level, it provides a more detailed analysis for the municipalities of Santa Cruz and Sucre. In the case of Bolivia, data is only available by departments so there is no census information disaggregated at the municipality level. Nonetheless, the solutions adopted – survey fieldwork and Delphi Method – bear testimony to the flexibility of the proposed methodology and its suitability for different scenarios and circumstances. The results show high levels of both vulnerability and industrial hazards, especially in the departments with highest economic development. The chapter also draws attention to the need for developing urban planning actions oriented towards a positive evolution of the management of these hazards.

Chapter 6 presents the results of mapping industrial risk in Argentina and explains the empirical procedures followed to obtain the maps. The chapter is divided into three parts. The first part provides background information about industrial hazardousness in Argentina. The second part studies the distribution of risk in the country, using the department or municipality as the unit of analysis. The third part presents a case study of the region with the highest concentration of departments/municipalities at high risk: the MABA (Metropolitan Area of Buenos Aires) using the census block group or census unit[3] as the unit of analysis. The chapter also explores qualitatively situations of 'environmental injustice' and notes that the conclusions regarding the correlation between vulnerability and environmental hazard in the case study differ from those obtained at national level. When the unit of analysis is 'census unit' group the spatial distribution suggests an inverse relationship between vulnerability and environmental hazard, where the risk gradient decreases with distance from the city of Buenos Aires as the social gradient of vulnerability increases. Although more detailed studies are required, this result suggests the need to develop indicators including different geographical units of analysis to examine local changes in the distribution of hazard trends.

Chapter 7 presents the evaluation of environmental industrial risk in Spain and describes social vulnerabilities and industrial hazard in a Spanish context following the methodology outlined in previous chapters. The empirical data collected allowed us to calculate a risk index for the whole of Spain and to develop more detailed spatial analysis at the level of the census unit for two specific case studies; the cities of Madrid and Seville. The risk index at the national level identifies the Spanish towns at greatest risk from the combined factors of social vulnerability and industrial hazardousness, while the case studies' findings show that there is no 'hot-spot' in Madrid but certain areas of Seville are exposed to a very high combined risk.

Finally, Chapter 8 provides a summary of the conclusions drawn from all previous chapters discusses its implications and provides policy recommendations. Limitations are stated as well as directions for further research suggested.

References

Adger, W. (2006). Vulnerability. *Global Environmental Change, 16*, 268–281.
Arenal Moya, C. (2009). *La Comunidad Iberoamericana de Naciones*, Laboratorio Iberoamericano. Documento de Trabajo 2009/1.
Baker, S. (2006). *Sustainable development*. London: Routledge.
Catan, T., & Lyons, J. (2008). World news: Spain's bets sour in Latin America. *Wall Street Journal*.
Clark, W., & Dickson, N. (2003). Sustainability science: The emerging research program. *Proceedings of the National Academy of Sciences, 100*(14), 8059–8061.
Dasgupta, S., Lucas, E. B., & Wheeler, D. (1998). Small plants, industrial pollution and poverty, evidence from Brazil and Mexico, *Working Paper 2029*. Washington, DC: World Bank Development Research Group.
Gray, L., & Moseley, W. (2005). A geographical perspective on poverty–environment interactions. *Geographical Journal, 171*(1), 9–23.

---

[3] A census unit is the smallest data for which census data is collected in a country, typically contain between 0 and 1000 people and up to 250 housing units.

Hart, S. L. (1995). A natural-resource-based view of the firm. *Academy of Management Review, 20*, 986–1014.
Hettige, M., Mani, M., & Wheeler, D. (1998). *Pollution intensity in economic development. Kuznet revisited*. Washington, DC. *World Bank Development Research Working Paper*, 1876.
Jones, B. (2001). Investing in Latin America: Spain and Portugal look to their former colonies. *Europe, 408*, 15–16.
Kates, R., Clark, W., Corell, R., Hall, J., Jaeger, C., Lowe, I., et al. (2001). Sustainability science. *Science, 292*(5517), 641–642.
Kates, R. W., & Dasgupta P. (2007). African poverty: A grand challenge for sustainability science. *Proceeding of the Natural Academy of Sciences USA, 104*, 16747–16750.
Kua, H. W., & Ashford, N. (2004). Co-optimization through increasing willingness, opportunity and capacity: A generalisable concept of appropriate technology transfer. *International Journal for Technology Transfer and Commercialization, 3*(3), 324–334.
Malamud, C., Steinberg, F., & Tejedor, C. (2011). *Anuario Iberoamericano 2011*. Real Instituto Elcano y Agencia EFE, S.A.
Melé, D., Debeljuh, P., & Arruda, M. C. (2006). Corporate ethical policies in large corporations in Argentina, Brazil and Spain. *Journal of Business Ethics, 63*, 21–38.
Moffet, M., & Prada, P. (2010). Latin America Frets While Spain Burns, *Wall Street Journal (online)*, Accessed April 2010.
Palmer, M., Bernhardt, E., Chornesky, E., Collins, S., Dobson, A., Duke, C., et al. (2005). Ecological science and sustainability for the 21st century. *Frontiers in Ecology and the Environment, 3*(1), 4–11.
Pargal, S., & Wheeler, D. (1996). Informal regulation of industrial pollution in developing countries: Evidence from Indonesia. *The Journal of Political Economy, 104*, 1314–1327.
Platt, G. (2000). Spain discover Latin America. *Global Finance, 14*(9), 26.
Quindimil Lopez, J. (2010). El Desarrollo en la Comunidad Iberoamericana de Naciones. *Laboratorio Iberoamericano*. Documento de Trabajo 2010/2.
Sachs, J. D. (2004). Sustainable development. *Science, 304*, 649.
Swart, R. J., Raskin, P., & Ribonson, J. (2004). The problem of the future: sustainability science and scenario analysis. *Global Change Environment, 14*, 137–146.
Taylor, M. (2008). Critical transport infrastructure in urban areas: Impacts of traffic incidents assessed using accessibility based network vulnerability analysis. *Growth and Change, 39*(4), 593–616.
Trelles Soliz, E., & Quiroz, C. (1997). Conclusiones Grupo IV: Consumo, derechos humanos, riesgo y educación ambiental. *II Congreso Iberoamericano de Educación Ambiental*, Guadalajara.
Valdaliso, J. M. (2008). Empresas y Grupos Empresariales en América Latina, España y Portugal. *Business History Review, 82*(1), 162–164.
Vazquez-Brust, D. A., & Liston-Heyes, C. (2010). Environmental management intentions: An empirical investigation of Argentina's polluting firms. *Journal of Environmental Management, 91*, 1111–1122.

Chapter 2
Evaluating the Firm's Environmental Risk: A Conceptual Framework

Diego A. Vazquez-Brust, Claudia E. Natenzon, Jerónimo de Burgos-Jiménez, José A. Plaza-Úbeda, and Sergio D. López

Abstract This chapter is organised into three sections. The first two sections outline the conceptual framework and the empirical procedure used in this project to evaluate the environmental risk generated by the firm. Risk is defined as the result of combining potential hazard, vulnerability and exposure. The framework suggests that the gap separating real – or managed – risk and potential – or evaluated – risk widens with less uncertainty and greater governability. The third section provides a description of the governability of environmental impact in Latin America, where the divide between real and potential risk is low, and where, therefore, methodologies that evaluate potential risk may also be appropriate for interpreting the extent to which evaluated risk is being managed.

Keywords Environmental risk · Uncertainty · Social vulnerability · Governability · CSR in Latin-America

1 Introduction and Conceptual Model

The concept of risk has been incorporated into our everyday routines and discourse. In its simplest definition risk is the likelihood of something happening. Academic definitions of risk are manifold and vary according to the scientific discipline and the analysis perspective. In economics risk is defined as the quantifiable likelihood that a (potentially damaging) phenomenon will occur. When this likelihood cannot be determined with any certainty we find ourselves in a situation of uncertainty. Starting from this definition, Downing (2001) and other authors define risk as the likelihood that a threat or potentially damaging phenomenon will take place. Other authors consider that this definition is limited by quantitative and deterministic notions of 'normal science', warning that its use in evaluating

---

Due to the differential characteristics of governability of the environmental impact in Spain, where the regulatory, social and corporate context is markedly different from other Ibero-American states, this chapter does not analyse the Spanish situation, which will be dealt with in Chapter 7.

D.A. Vazquez-Brust (✉)
The Centre for Business Relationships, Accountability, Sustainability and Society (BRASS), Cardiff University, Cardiff Wales CF10 3AT, UK
e-mail: VazquezD@cardiff.ac.uk

risk may lead us to underestimate or even ignore the weight of cultural, social and political aspects that are difficult to quantify but on which decisions must be taken (Renn 1992).[1]

The present work uses a definition of risk based on social theories developed by sociologists such as Giddens (1990), Beck (1992), Marris (1996) and Crichton (1999). For Crichton (1999) the risk that a potentially harmful phenomenon (threat) may have damaging results for a given social system depends on three independent but intertwined factors or dimensions: the hazardousness of the threat, the degree to which the social system is exposed to the threat and the social vulnerability of the system, i.e. how prone to damage the social system is when it is in danger.

Within this framework we propose that a situation of evaluated (potential) risk is preceded by state of potential hazard, which we define as a magnitude of the greatest overall adverse impact that a threat can generate (hazardousness), and which depends on the nature of the threat to which the system is exposed (for instance the length of time a flood will affect the social system or the likelihood of it recurring). Likewise, the evaluation of a situation of potential risk precedes the accident or disaster. This evaluated risk emerges from the potential danger of a phenomenon or process which threatens a given social group. The degree to which each social group may be injured or damaged corresponds to its vulnerability and constitutes the second dimension of risk.

In order to know how this risk becomes damage to the system we propose incorporating two additional dimensions: uncertainty and governability. Uncertainty refers to aspects of a problem that are unknown but on which decisions must nevertheless be taken (Giddens 1991, 1992; Funtowicz and Ravetz 1993). The fifth dimension of risk proposed in our framework is 'governability'. Marris (1996) states that the governability of risk is of particular relevance for the analysis of technological threats, i.e. those generated by human activity. Governability refers to external institutional and cultural factors that act as a 'buffer', moderating the other dimensions of risk. By way of example, the existence of regulations or communities that are capable of mobilising protest increases the governability of risk, while the opposite occurs when control is lacking or when there are implicit social conventions that normalise deviant corporate practices (i.e. when non-compliance of environmental regulation is tolerated by communities).

Figure 2.1 transfers the theoretical framework to a conceptual model:

---

[1] Otwin Renn (1992) identifies seven approaches to the concept and evaluation of risk: (1) Actuarial Approach: Application of the calculation of probabilities, statistics and financial mathematics to predictions of risk and insurance, (2) Toxicological and epidemiological approach (including ecotoxicology), (3) Engineering approach (including probabilistic risk evaluation), (4) Economic approach (including risk-profit comparison), (5) Psychological approach (including psychometric analysis), (6) Social Theories of Risk, and (7) Cultural Theory of Risk (using 'gris-group' analysis).

2 Evaluating the Firm's Environmental Risk: A Conceptual Framework

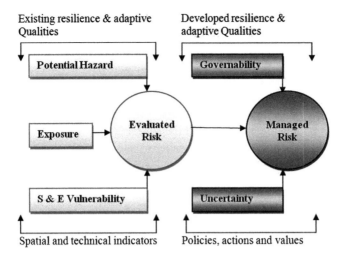

Fig. 2.1 Conceptual model: dimensions of risk

This book focuses on identifying evaluated risk by generating quantitative indicators for the social vulnerability and industrial hazardousness dimensions. A combined approach of qualitative and quantitative methodologies is necessary to incorporate the other dimensions as well as to estimate the divide between evaluated/potential and managed/real risk and its implications for decision making. The following section analyses the main dimensions that appear in Fig. 2.1 and how they are applied in cases of risk generated by industrial dangers.

1.1 Hazardousness

In this work industrial activity is considered to produce hazards or phenomena that have potentially damaging effects. Industrial activity implies carrying out tasks or handling substances that are a potential danger for the social and environmental system in which the activity is located. In other words, industry involves tasks that have the potential to generate negative impacts. Potentially damaging substances must not only be manipulated, they must also be transported to the industrial sites, and once they have been transformed they are once again transported to their final destination, either as products for consumption or as waste material of the manufacturing process. In addition, this process generates different compounds that are often released into the surrounding environment, such as gases, vapours, solids, sewage, etc.

The magnitude of the greatest overall negative impact that the firm may generate is termed its hazardousness or perilousness. This hazardousness will be greater the larger the firm, the more dangerous the substances they process, the more waste they generate, the greater the density of firms per square kilometre, etc. Typical examples of industries, with a high degree of hazardousness, are the nuclear industry, chemical and pharmaceutical industries, the steel industry and mining (Perrow 1984).

1.2 Exposure

Exposure refers to the degree to which the population, property and goods could be materially affected by a dangerous phenomenon. It is a consequence of the relationship between hazard and vulnerability, and in turn it influences both. This dimension is reflected in space by a historical construct that unifies natural processes and socio-economic relationships. Thus, certain land uses and distribution of wealth and population are generated (Natenzon 2003). Bearing in mind industrial activity, the risk increases when potentially dangerous industries are located in urban areas, in close contact with residential areas, housing and educational, health and administrative installations. In such a scenario, there is a high likelihood that the inhabitants of these areas will suffer. Greater proximity of social systems to industrial activities increases the likelihood of suffering economic, environmental and human damage or losses as a result of the deficient or accidental functioning of the technology applied in an industry (Edelstein 1987).

1.3 Vulnerability

The United Nations/International Strategy for Disaster Reduction (UN/ISDR 2004) defines vulnerability as 'the conditions determined by physical, social, economic and environmental factors or processes which increase the susceptibility of a community to the impact of hazards'.

The concept of vulnerability is considered one of the strongest analytical tools to describe states of susceptibility to damage of social and physical systems, to identify situations of lack of power and marginality and to guide regulatory analyses of actions intended to improve standards of living by reducing risk (Adger et al. 2004). This concept is used to analyse risks due to climate change, and Social Vulnerability is an intrinsic property of human systems that makes certain communities more prone to suffering economic losses or the loss of human lives as a consequence of environmental hazards in the form of shocks, e.g. floods caused by climate change, or in the form of stress, as in the case of exposure to contaminated rivers (Adger et al. 2004). Cardona (2005) relate social vulnerability to socioeconomic fragility (poverty, inequality, unemployment and debt, access to food, insurance and credit) and the lack of resistance or ability to withstand negative impacts (marginality, access to education, quality of housing and infrastructure of services and health, life expectancy or social security).

Thus, authors studying poverty and development define a state of vulnerability to poverty in the face of economic risk (Morduch 1994; Yapa 2002; Scott 2006). These authors see poverty as a complex phenomenon whose prevention must integrate the minimisation of social vulnerabilities, but also of environmental ones.[2]

---

[2] As an example, we can take Yapa (2002)'s work, who documented the social and environmental consequences of the introduction of genetically modified rice in Sri Lanka. Yapa argues that the

Siegel and Alwang (1999) identify a vicious circle of vulnerability in which environmental vulnerability leads to less economic productivity and greater vulnerability to poverty. Areas with low per capita net product and low density are more vulnerable to economic and environmental risks (Dilley et al. 2005). Scott (2006) considers that social vulnerability and environmental risk are two of the key factors in generating poverty, while Hart (1995) postulates that environmental deterioration occurs as a consequence of poverty, technology and population increase.

Our theoretical framework considers that social vulnerability includes all the characteristics of a system that are independent of the hazards to which it is exposed but which influence the final result of a hazardous event (Natenzon 2003; Allen 2003; Adger et al. 2004; Gallopin 2006). Some Latin-American authors (Minujín 1999; Filgueira and Peri 2004) have proposed the use of this notion to explore the social consequences of socio-economic processes occurring in the last two decades that traditional dichotomous concepts of poverty – wealth have not acknowledged (for instance exclusion from decision processes, lack of power within local communities, unequal access and benefit sharing of ecosystems services – clean water, air, land). It should be stressed that the concept of social vulnerability differs from the perspective used in natural sciences, which uses the term vulnerability as 'biophysical vulnerability' of a system and understands that it is a function of the intensity and likelihood that a hazard will occur, and it depends on the magnitude of the final damage (lives lost, material damage) that a system might suffer after a disaster.[3]

1.4 Governability

However, although any industrial activity involves a certain degree of hazard, and although there are vulnerable social systems exposed to these potential dangers, it is not inevitable that the potential impact will cause damage to the environment or to people, as the impact or hazard of the industry will be measured by the conditions of governability in which the industrial activity is carried out, and which act as a dissuasive element, in order that firms avoid contaminating or at least implant measures to reduce the likelihood that its activities may have negative

---

increase in expenses due to the use of pesticides and to treating the diseases they provoked actually made the farmers poorer, as well as degrading their means of subsistence and eclipsing cheaper and more sustainable crop techniques.

[3] Adger et al. (2004) define biophysical vulnerability as the result of combining four factors: (a) the nature of the hazard to which the system is exposed (for instance the duration of a flood or the likelihood of it recurring), (b) the probability of a hazard occurring, (c) the degree of exposure to the hazard (d) the intrinsic sensitivity or incapacity of the system to resist the adverse effects of the hazard to which it is exposed (this is equivalent to the concept of social vulnerability). Yapa (2002) makes the distinction between intrinsic social vulnerability, which is not a function of the hazard to which the system is exposed, and relative social vulnerability, which refers to the characteristics of a system that make it more vulnerable to certain types of hazard. For instance, the construction of housing below the flood elevation in areas susceptible to flooding increases the vulnerability to flooding, but not to industrial atmospheric pollution.

impacts on the environment or people (Becker 1968). These conditions of governability comprise three factors: state, market and community governability (Afsah et al. 1996). State governability refers to the existence of sound environmental regulation and suitable enforcement (Heyes 1998). Market governability is linked to the effect of environmental damage on the firm's reputation, the existence of a price structure and a consumer sector that rewards 'clean' firms, and financial institutions, insurance companies or investors that penalise polluting firms (Siegel 2009). Finally, community governability includes pressure on the part of the media and communities (Hasnas 1998; Deegan 2002), social norms and customs that dictate which behaviour is acceptable (Dryzek 1997) and genuine voluntary or philanthropic action by the firm to improve its environmental performance (Scherer and Palazzo 2007; Aguilera et al. 2008).

When conditions of governability exist, the likelihood of situations of 'environmental stress' due to constant emissions of polluting substances is reduced considerably (Wheeler 2004). European nations (excluding Eastern Europe), the USA, Japan and Canada are deemed to be countries with high levels of governability (Jagnicke 1985; Erickson 1994). Nonetheless, catastrophes such as the nuclear explosion at Three Mile Island (USA) and the release of carcinogenic vapour clouds in Seveso (Italy) have shown beyond doubt that high levels of governability are no guarantee against such damage. A degree of hazardousness will always exist where industry is present, and the possibility of an environmental catastrophe cannot be discarded, whether it be due to human error, unforeseen or underestimated technological problems, external factors or uncontrollable circumstances such as natural disasters or war (Beck 1992; Erickson 1994).

1.5 Uncertainty

Uncertainty is inherent to the scientific process and it may refer to the general state of knowledge of a problem, to the limits of the paradigm used for its analysis, to the methodological quality of existing information or to the ambiguity of interpretation of certain data (see, for example, Leach et al. 2007 and Scoones et al. 2007 for an analysis of differences between uncertainty, ignorance and ambiguity in aspects related to risk). In all cases it refers to aspects for which certain knowledge is not available, but on which one must nevertheless take decisions (Funtowicz and Ravetz 1993). From the scientific point of view, one speaks of risk when one can quantify or establish a probability. When the probability of what may happen, albeit an approximate estimate cannot be established, we are not dealing with risk, but rather with *uncertainty* and the future is unclear.

From a technical point of view, whereas risk implies knowledge, uncertainty implies insecurity due to lack of knowledge or due to the complexity/instability of the empirical system studied (Wynne 1992; López Cerezo and Luján-Lopez 2000). Uncertainty arises from relative ignorance of hazard, vulnerability, exposure and governability. If one detects uncertainty from the outset, it can be dealt with and, therefore, be included as a constituent dimension of risk.

1.6 Adaptation and Resilience as System Qualities

Similarly to the concept of vulnerability, resilience and adaptation – as qualities of a system – also have nuances that are different in natural sciences and social sciences (Wisner et al. 2004; Adger et al. 2004).

Resilience refers to intrinsic capabilities of a living creature or ecosystem that allows it to face a threat or danger without suffering damage or substantial changes in structure and function, and even return to the situation before the hazardous event. It has therefore been taken as synonymous with 'elasticity.' However, we know that social configurations never return to a previous configuration, the emerging configurations of history are always different. Therefore the reversibility implicit in the physical and even ecological systems does not corresponds in social systems: After a historic event, emerging social configurations will never be the same. 'Intrinsic' characteristics of an individual or social group (capabilities) are the result of a previous history, and are not the only ones that come into play when coping with dangers: context characteristics are equally or even more relevant than the capabilities of their own.

In the field of risk analysis, the word resilience (as we can see, a notion with major ambiguities) seems to be referring to what has traditionally been called 'preparation', which connotes a deliberate willingness to take into account the existing risk in order to develop a managed risk. In this way, it is necessary to look at both capabilities and vulnerabilities; and what to do to improve the first and to decrease the latter.

From the standpoint of natural sciences, adaptation means maintaining life, but this definition does not suffice for social sciences. It does not imply resignation, but rather maintaining life in worthy conditions, which depends on factors linked to the development process and the democratic quality of the social system (Vazquez-Brust et al. 2009). Adaptation can be reinforced by adequate planning, but it requires the social system to have the capacity to adapt. The adaptive capacity of a social system can be defined as its ability to plan and implement adaptation processes, or more generally as the capacity of a human system to modify itself in order to increase, or at least maintain, the standard of living of its members in the face a range of current or future disturbances to the physical or social environment (Gallopin 2006).

The governability of a system also influences the management of risk through actions and policies that increase adaptation, i.e. that help the social system to adapt to an evaluated risk by reducing the likelihood of an event or mitigating its damaging consequences, or by accelerating the recovery and response of the social system affected (Yapa 2002; Du Toit 2004; Eakin and Lemos 2006; Scott 2006; Archer et al. 2006). The adaptive capacity of human systems depends to a great extent on coordinated collective and institutional actions that are facilitated when conditions of governability exist. Community governability contributes to increasing adaptive capacity by developing social capital: mutual trust, social integration, community networks, norms, consensus and the flow of information used by individuals for their own benefit or for that of the community. Other factors that influence adaptive capacity, such as level of income, saving capacity, technology and infrastructure, know-how and skills, equality, quality and power of the institutions, credit access, insurance and employment, are closely linked to state and market governability (Yapa 2002; Eakin and Lemos 2006).

How do we differentiate resilience from adaptation? Adaptation is concerned with the specific responses/policies that operationalise resilience in each particular case. Thus, resilience refers to the intrinsic structure of each social group whereas adaptation/adaptive capability refers to the dynamics of interaction between social groups and its context. Adaptation brings different aspects of resilience into play depending on the type of danger faced by the social group, since the factors from which systems build their adaptive capacities are different for different threats.

The practical application of the conceptual elements outlined has been implemented in the project by means of a quantitative procedure for calculating indicators of evaluated risk, which is described in the following section.

2 Empirical Procedure for Risk Assessment

Evaluated risk is measured in this Project as the result of an empirical procedure combining quantitative indices of vulnerability and industrial hazard in a geographic unit of analysis. The values of such indices had been obtained by statistical procedures through which we classified indicators of social vulnerability and industrial hazards into discrete ranges or classes associated with a qualitative level (1 = very low to 5 = very high). The procedures and data used to calculate the indices of social vulnerability and hazard are described in Chapters 3 and 4 respectively.

To create the indicator of evaluated risk we related each of the ranges of perilousness with each of the ranges of vulnerability using a very simple procedure, shown in Table 2.1, which could be called 'multi-criteria' – to combine variables of

Table 2.1 Evaluated industrial risk: the combination of hazardousness index with vulnerability index

Hazardousness/ Vulnerability	1. Very Low	2. Low	3. Medium	4. High	5. Very High
1. Very Low	1+1=2	1+2=3	1+3=4	1+4=5	1+5=6
2. Low	2+1=3	2+2=4	2+3=5	2+4=6	2+5=7
3. Medium	3+1=4	3+2=5	3+3=6	3+4=7	3+5=8
4. High	4+1=5	4+2=6	4+3=7	4+4=8	4+5=9
5. Very High	5+1=6	5+2=7	5+3=8	4+5=9	5+5=10

Table 2.2 Ranges of evaluated industrial risk

Level of Risk	Range and colour
Very low	2–3
Low	4–5
Medium	6
High	7–8
Very high	9–10

different nature. The 25 possible combinations of evaluated risk values thus obtained (minimum value 1 + 1 = 2, maximum 5 + 5 = 10) were reclassified into five ranges, from very low to very high risk (Table 2.2).

Once the elements of our theoretical framework and risk assessment procedure have been defined, the following section makes a qualitative, desk-research-based analysis of the dimension of governability of environmental impacts generated by firms in Latin America. As we have seen, this analysis is essential to appreciate how evaluated risk (the result of this Project) differs from managed risk.

3 Governability of Industrial Risk in Latin America

Table 2.3 provides information on CO_2 emissions and the number of firms with ISO 14001 certification in Ibero-America.[4] These indicators allow a cursory cross-country comparison of environmental deterioration and industry efforts to improve environmental governance. The environmental indicators are complemented with basic contextual information about GDP, Urban Poverty, Extreme Poverty[5] and Distribution of Income (GINI coefficient)[6] in each country (Chapter 3 presents a more detailed and nuanced analysis of social indicators in Latin-America).

A first observation of Table 2.3 reveals quantitative differences in the patterns of poverty distribution in the region. In this regard, Latin American Countries can be classified into four groups. A small group of countries has levels of poverty below 20% and indigence below 5% (The 'ConoSur' countries: Argentina, Uruguay,

---

[4] Number of firms implementing ISO 14001 standard is often used as an indicator of voluntary environmental responsibility, therefore the more firms implementing the standard, the higher the governability of a country. CO_2 emissions data used here represent the mass of CO_2, a potent greenhouse gas, produced during the combustion of solid, liquid, and gaseous fuels, as well as from the manufacture of cement (CO_2 is produced as a byproduct as cement is calcined to produce calcium oxide) and gas flaring. Environmentally aware industries will endeavour to reduce their CO_2 emissions per unit of GDP.

[5] People in situation of poverty live in households with an aggregated income which is not enough to meet their basic food and non-food needs (housing, education, health). People living in extreme poverty or indigents are defined as persons whose household has an income so low that they cannot buy enough food to adequately cover their nutritional needs (ECLAC 2010).

[6] This is the most commonly used measure of inequality. The coefficient varies between 0, which reflects complete equality and 1, which indicates complete inequality (one person has all the income or consumption, all others have none).

Table 2.3 Poverty and environmental deterioration in Latin-America

	Gross domestic product			Poverty and distribution[a]						Environmental impact			
	GDP/ capita 2007	GDP 2007	GDP 2009	Total poverty (%) 2009	Urban poverty (%) 2009	Extreme poverty (%) 2009	Extreme urban poverty (%) 2009	Urban GINI coef 2009	CO_2 emm/capita 2007	CO_2 tot (ton) 2007	Number ISO 2007	Variation number ISO 2009 2007
Argentina	9353.5	374444.4	397647.1	15.8.[a]	11.3	4.4	3.8	0.51	4.6	183728	676	−335
Belize	3883.6	3333.3	1176.5	S/D	S/D	S/D	S/D	S/D	1.4	425	4	3
Bolivia	1125.0	35555.6	11851.9	54.0	42.4	31.2	16.2	0.5	1.4	13190	32	0
Brazil	4290.8	624000.0	829375.0	24.9	22.1	7.0	5.5	0.56	1.9	368317	1327	−545
Chile	6100.4	213913.0	102857.1	11.5	11.7	3.6	3.5	0.52	4.3	71705	576	84
Colombia	3067.8	64375.0	139756.1	45.7	39.7	16.5	12.4	0.55	1.4	63439	573	264
Costa Rica	5131.9	43913.0	23076.9	18.9	18.5	6.9	5.4	0.49	1.8	81119	90	−11
Cuba	4190.4	1590.9	48000.0	S/D	S/D	S/D	S/D	S/D	2.4	27055	24	17
Dominican Rep.	3553.0	30000.0	35714.3	41.1	39.3	21	19.4	0.58	2.2	20759	25	13
Ecuador	1679.3	22285.7	23913.0	42.2	40.2	18.1	15.5	0.48	2.2	29989	110	32
El Salvador	2620.5	16000.0	15000.0	47.9	42.3	17.3	12.8	0.45	1.1	6700	12	4
Guatemala	1614.5	21428.6	25000.0	54.8	42.0	29.1	14.8	0.54	1	12930	15	0
Honduras	1423.4	10000.0	10416.7	68.9	56.9	45.6	26.2	0.49	1.2	8834	25	7
Mexico	7037.6	739000.0	725000.0	34.8	29.3	11.2	6.4	0.49	4.4	471459	870	131
Nicaragua	881.7	5000.0	5000.0	69.3	63.8	42.4	33.4	0.5	0.8	4591	5	3
Panama	5222.7	17222.2	1857.1	26.4	16.3	11.1	4.6	0.47	2.2	7250	13	−18
Paraguay	1463.9	8571.4	9000.0	56.0	48.2	30.4	19.0	0.43	0.7	4133	9	3
Peru	2694.4	76000.0	83809.5	34.8	31.1	11.5	2.8	0.42	1.5	42988	176	62
Uruguay	7426.0	25217.4	27307.7	10.7	10.4	2.0	2.0	0.43	1.9	6219	71	13
Venezuela	5602.9	163333.3	165000.0	27.6		9.9			6	165550	99	50
Spain	16200.0	746820.0	751706.6	19.6[b]	14.3	28.5	28.5	0.2	8	358965.7		
Portugal	11220.0	119156.4	121418.46	22.3[b]				0.4	5.5	58063		

[a] INDEC: Instituto Nacional de Estadisticas Argentina
[b] Panel A. Percentage of persons living with less than 50% of median equivalised household income, late-2000s. *Source*: EN (2009) at http://www.tradingeconomics.com/spain/indicators

Source: ECLAC (2010)

Chile); these are closely followed by Costa Rica and Brazil (both with poverty below 25% and indigence below 10%). On the other extreme we find a group of extremely disadvantaged countries with more than 50% of poor and more than 25% of indigents (Bolivia, Nicaragua, Honduras, Guatemala and Paraguay) and a majority of countries – including Mexico and Venezuela – with poverty levels between 25 and 50% and indigence between 10 and 25%The observation of GINI coefficients in Table 2.3 indicates a great level of inequality in all Latin American nations, Brazil being the country with the greatest inequality in the distribution of wealth. Inequality creates a powerful disincentive to respond effectively to the need to manage an evaluated risk. High inequality harms social cohesion, which is the basis of social capital and a key factor in the development of adaptive and resilience qualities in a social system. In particular, lack of social cohesion hampers the success of strategies and actions requiring multi-stakeholders engagement and bottom-up governance (Vazquez-Brust et al. 2009).

Although internal migrations and immigrants contributed to a sharp rise of urban poverty during the 90s, the percentage of people living in poverty is still significantly higher in rural populations than in urban dwellings. Nowadays, 27.8% of total population in urban areas in Latin-America lives in poverty and 8.8% in indigence, while a staggering 52.8% of total population in rural areas lives in poverty and 30% of them are indigents (ECLAC 2010). However, in the last three decades Latin America has tripled its urban population, which is now estimated to reach 79.1% (471 millions) of the region's total inhabitants. Thus, in absolute terms the number of urban poor is more than twice the number of rural poor (ECLAC 2010). Additionally the poor and indigents live that live in urban areas tend to be more exposed to environmental threats generated by industrial activities and suffer more environmental illnesses than the poor in rural areas.

The observation of Table 2.3 reveals that Latin-American economies have been unable to decouple environmental damage from economic growth. Indeed environmental deterioration environmental goes hand in hand with GDP. The 4 Latin-American countries with higher GDP also produced two thirds of total emissions in the area: Argentina, Brazil, Venezuela and Mexico. The same countries accounted for three quarters of all ISO-14001 certified enterprises in the region.[7] Alarmingly, the total number of ISO certified firms dropped from 2007 to 2009, despite widespread constant increase of GDP and CO_2 emissions. The increase in certified firms in Mexico and Venezuela did not compensate the substantial drop in numbers of certified firms in Argentina (33% less than in 2007) and Brazil (29% less than in 2009). Argentina and Brazil were pioneers in the promotion of ISO 14001 certification. From 2003 to 2007 the numbers of certified firms in these countries had grown steadily. However, the sharp decline in certification experienced in the last three years, hints that firms no longer are willing to voluntarily improve their

---

[7] ISO 14001 is a voluntary standard for environmental management based on principles of compliance with national legislation and continuous improvement. Companies with ISO 14001 certification use their own management systems but must have external audits to assess their environment performance.

environmental performance. Thus, conditions of governability are weakened and the likelihood of situations of 'environmental stress' due to constant emissions of polluting substances is increased considerably.

In this context, analysis of pathways to achieve corporate excellence in environmental aspects is extremely relevant for Latin America, both from the firm's viewpoint and from the perspective of society as a whole (Peinado-Vara 2006). As far as the firm is concerned, Latin American firms that export to markets in the North struggle to combat the prejudice that production in the South is 'dirty', and this constitutes a strategic disadvantage to access markets where consumers attach great importance to environmental concerns (Schmidheiny 2006). As for the social viewpoint, firms' environmental performance is fundamental for the welfare of communities in Latin America (Haslam 2004; Peinado-Vara 2006). Indeed, the evidence indicates that firms in Latin America tend to cause greater environmental damage than their counterparts in Europe or North America (Dasgupta et al. 2000; Pratt and Fintel 2002 and Ruiz-Tagle 2003). For decades Latin American governments have fomented models of substitution of imports with protected markets that have given rise to production methods that are, generally speaking, inefficient, with high levels of contamination and high consumption of non-renewable resources (Pratt and Fintel 2002). This production system has even had negative consequences for the health of factory workers and has threatened the surrounding population with possible industrial accidents (Porto de Souza and Freitas 2003).

On the whole, legislation and regulations have done little to improve firms' environmental performance (Ruiz-Tagle 2003). Regarding public policies of the State, when firms generate high levels of accidents and illness, they are neither sanctioned, nor do they receive incentives to invest in preventing future accidents and thus improve performance (Porto de Souza 2007:6). Although most Latin American countries currently have some form of environmental regulation, it tends to be inefficient in softening the impact of industrially generated pollution (Eskeland and Jimenez 1992). The regulation's lack of effectiveness is sometimes due to design faults, incorporating norms that simply copy European or North American laws, making no effort to adapt them to the Latin American context (Pratt and Fintel 2002). Nonetheless, the most important factor is related to problems of governability and the fact that the institutions responsible for enforcing the legislation are not up to the task. The lack of resources to ensure that firms meet environmental regulations is a recurring theme (Vazquez-Brust et al. 2010). There is a lack not only of human and economic resources to inspect firms, but also of methodological tools to identify priority areas of intervention, and this is compounded by a lack of political will to penalise polluting firms (Ruiz-Tagle 2003; Birdsall and Wheeler 1992). This is due in part to corruption (Guidi 2008), and in part to the attitude that contamination is an acceptable price to pay for an industry that generates economic development (Dasgupta et al. 2000).

Although the growing deterioration of the environment in suburban areas has increased overall environmental awareness (Dasgupta et al. 2000; Pratt and Fintel 2002), civil society and NGOs have not been powerful enough to make up for the deficient regulation and to oblige polluting firms to improve their environmental performance (Birdsall and Wheeler 1992; Guidi 2008). This is because potentially contaminating or 'dirty' industries tend to be situated close to the poorest communities, which on the whole are poorly informed of the consequences of contamination, are less concerned about environmental quality and are afraid of losing jobs if contaminating firms are penalised (Dasgupta and Wheeler 2001; Grynspan and Kliksberg 2008; Vazquez-Brust et al. 2009). Even when they are aware of the problem and are prepared to press for their right to live in a clean environment, they do not possess the resources or access to the institutions that are necessary to make their claims heard (Martínez-Alier 2002) and achieve environmental justice (Porto de Souza 2007).

Researchers in the field of corporate social responsibility argue that Latin American subsidiaries of multinational firms are already under pressure from their parent companies to be environmentally responsible and to comply with global standards (ISO, Global Compact). This is partly due to the growing demand for green products at the end of the supply chain (Torres-Baumgarten and Yucetepe 2009; Prieto-Carron et al. 2006), and partly due to the moral pressure received from insurance companies and investment funds. On the other hand, local firms that export their goods to the USA and Europe have to adapt to the requirements of their customers regarding environmental behaviour (Schmidheiny 2006). Finally, small and medium-sized firms have more reasons to respect the environment than large firms, since they are local concerns whose owners, managers and workers form an integral part of the communities where they operate (Quinn 1997). In Latin America small firms tend to seek 'satisfactory benefits' rather than maximised profits. Their owners and management may be more willing to forego some profit if it means increasing their self-satisfaction in producing goods that they like, helping the most vulnerable in society or giving something back to their communities (Vives 2006).

The findings of some case studies of firms that are taking a leading role in solving environmental problems even though they are under no socio-institutional pressure, provides empirical support for those who claim that the way towards sustainable development in Latin America is through voluntary effort of Corporate Social Responsibility – CSR (Pratt and Fintel 2002; Vives 2006; Guidi 2008). Vives (2006) states that an emerging new generation of managers in Latin America is committed to integrating environmental considerations in their business strategies, on the one hand because they are ethically motivated to assume social and environmental responsibilities, and on the other because they take measures in anticipation of social and market pressure. Nonetheless, most of the results of empirical studies on the extent and priorities of CSR in Latin America portray a less promising scenario in terms of firms' environmental commitment (Vazquez-Brust and Liston-Heyes 2008). This geographical region has always been more concerned with social issues than with environmental ones (Peinado-Vara 2006), and philanthropy continues to be a more attractive channel for social commitment than the complex task of integrating environmental responsibilities and development into the firm's daily routine (Newell and Muro 2006). Although contaminating firms tend to reveal more information on social and environmental practices than their counterparts in sectors that are less environmentally sensitive (Araya 2006), there is still an alarmingly low level of responsibility regarding the environment among leading firms (Pratt and Fintel 2002). Empirical analyses suggest that programmes of environmental management and regeneration among multinational firms are extremely limited and of dubious efficacy (Torres-Baumgarten and Yucetepe 2009). The scenario is no better among small and medium-sized firms: in a survey financed by the Inter-American Development Bank in 6 Latin American countries, only one third of contaminating firms claimed to implement environmentally responsible practices (Vives 2006).

There may be several explanations as to why Latin American firms are reluctant to accept their environmental responsibilities. For local firms it is partly a question of lack of environmental values (Vives 2006). For small firms it is also a problem of resources (the financial resources of small and medium-sized firms are easily affected by recurring crises), as well as a problem of the communities that present their claims in a language that the firms cannot understand (Vives 2006) or with an attitude of confrontation that creates a barrier to cooperation (Vazquez-Brust et al. 2009).

In the case of multinationals, despite increasing knowledge and having good intentions regarding sustainability, local subsidiaries have not been able to translate these intentions into specific behaviours and results (Pumpim de Oliveira and Gardetti 2006). The literature puts forward many explanations for this, with varying degrees of complexity. Some suggest that the bottom line is simply that environmental responsibilities do not play a central role in larger companies (Chudnovsky et al. 2005; Peinado-Vara 2006). Indeed, the reduction in the numbers of ISO 14001 certified firms in pioneer Latin-American countries lends support to those arguing that ISO 14001 schemes are not adopted out of environmental responsibility, but to achieve benefits such as brand recognition and cost-savings. However, such benefits can be short-lived or just not achievable without substantial investments. Therefore, many companies leave the scheme disappointed because the environmental investment did not fulfil their expectations in terms of economic performance. In other cases, once low-hanging fruits (economic benefits achieved from low-cost environmental improvements) have been exhausted, firms abandon the standard because further improvements in environmental performance will require significant investment.

Other researchers point out a problem of lack of 'localised' environmental strategies. Managers in multinationals neither understand the needs of local communities nor are receptive to their demands (Newell and Muro 2006; Pumpim de Oliveira and Gardetti 2006), and by and large they have proved incapable of forging successful alliances with local organisations (Guidi 2008). As a result, most large firms have been unable to adapt the practices of environmental responsibility that originate from the parent company to the Latin American context (Haslam 2004). The practices employed have proved to be insufficient to obtain results in conditions of low governability, and management has not known how to tackle specific issues in developing countries, such as the combination of poverty and environmental deterioration that is characteristic of the urban landscape where firms operate in Latin America (Vazquez-Brust et al. 2009).

The outcome of all these institutional deficiencies is that the impact of industrial contamination is hardly softened, and the adaptive capacity of the communities is not fomented, with the result that 'potential hazard' effectively means the same as 'real hazard' and environmental degradation has increased (Hochstetler 2002; Pratt and Fintel 2002). This degradation in turn affects the welfare of communities, increasing the incidence of disease related to water and air pollution and reducing the reserves of natural, non-renewable resources such as drinking water (Dasgupta and Wheeler 2001; Guidi 2008). Grynspan and Kliksberg (2008:67) warn that if this situation is not remedied the ever greater environmental and social agenda due to industrial activity may endanger the dynamic of economic growth and development in Latin America. For inhabitants of urban conglomerations, the proximity of housing to industrial sites that handle dangerous substances in situations of low governability entails an obvious risk, whether it is due to an accident during the handling or transport of said substances, or the danger of suffering the effects of contamination caused by the high concentration of industrial sites (Porto de Souza 2007).

4 Implications of the Approach Adopted

The first part of the chapter provides a description of the conceptual framework used, while the second gives a brief description of the conditions of governability of environmental risk in Latin America, and argues that in this context, for a given condition of uncertainty, the divide separating the assessable risk of an activity and the real risk is narrow, mainly due to low governability. Therefore, the methodology for evaluating risk proposed in this work also represents a suitable preliminary tool for identifying manageable risk as a result of industrial activity.

The availability of methodologies for identifying areas of greater industrially-generated hazards constitutes a first step towards designing policies of intervention with a view to strengthening conditions of governability (Vazquez-Brust et al. 2009) and reducing uncertainty. Along these lines, in the course of this Project the availability of environmental indicators of easy access was studied in organisms linked to multinational institutions such as the Organisation of American States (OAS), the Economic Commission for Latin America and the Caribbean (ECLAC), the World Bank or the Inter-American Development Bank. The analysis shows very limited availability of indicators of industrially-generated environmental hazards. Furthermore, those that do exist, for instance the Millennium Development Indicators, are only partially available on a national scale. These indicators prove useful for identifying improvement or deterioration of the mean parameters of the country over time (e.g. carbon dioxide emissions). However, identifying situations of potential hazards requires indicators on a much more detailed scale that allow us to detect the problems suffered by those communities in highly industrial areas.

One direct implication of our conceptual approach is that the generation and diffusion of methodologies to evaluate risk and the dissemination of the products obtained in the communities affected constitutes action research, creating the potential to narrow the divide between evaluated risk and manageable risk. On the one hand, methodological advances to elaborate rigorous and replicable indicators and procedures to identify areas of risk contribute to reducing uncertainty. On the other hand, the diffusion of results among management groups related to the areas of risk, if accompanied by a process of debate on methodologies, their limitations and scope, contributes to the education and awareness of both those affected and those responsible, and creates the conditions for developing social processes that allow us not only to evaluate the risk, but also to manage it, thus fomenting governability.

Indicators for the analysis of social vulnerability are studied next in Chapter 3. In turn, the aim of Chapter 4 is to develop a methodology that allows us to detect the areas of greatest hazard, using techniques of spatial analysis to be found in Geographic Information Systems (GIS).

References

Adger, W., Brooks, N., Bentham, G., & Eriksen, S. (2004). *New indicators of vulnerability and adaptive capacity.* Tyndall Centre for Climate Change Research, Technical Report 7.

Afsah, S., Laplante, B., & Wheeler, D. (1996). *Controlling industrial pollution: A new paradigm.* Policy Research Department, World Bank: Working Paper N° 1672.

Aguilera, R., Rupp, D., Williams, C., & Ganapathi, J. (2008). Putting the S back in corporate social responsibility: A multi-level theory of social change in organizations. *Academic of Management Review, 32*(3), 836–863.

Allen, K. (2003). Vulnerability reduction and the community-based approach. In E. Pelling (Ed.), *Natural disasters and development in a globalising world* (pp. 170–184). London: Routledge.

Araya, M. (2006). Exploring Terra Incognita: Non-financial reporting in corporate Latin America. *Journal of Corporate Citizenship, 21,* 25–39.

Archer, D., Crocker, T., & Shogren, J. (2006). Choosing children's environmental risk. *Environment and Resource Economics, 33,* 347–369.

Beck, U. (1992). *Risk society. Toward a new modernity.* London: Sage.

Becker, G. S. (1968). Crime and punishment: An economic approach. *Journal of Political Economy, 76,* 169–217.

Birdsall, N., & Wheeler, D. (1992). Trade Policy and industrial pollution in Latin America: Where are the pollution heavens? In P. Low (Ed.), *International trade and the environment* (pp. 117–126). Washington DC: World Bank.

Cardona, O. D. (2005). *Indicators of disaster risk and risk management.* Summary Report. IDB/IDEA Program on Indicators for Disaster Risk Management. Inter-American Development Bank, Sustainable Development Department Environment Division: Washington, DC.

Chudnovsky, D., Pupato, G., & Gutman, V. (2005). *Environmental management and innovation in Argentine industry: Determinants and policy implications.* Buenos Aires: CENIT, mimeo.

Crichton, D. (1999). The risk triangle. In J. Ingleton (Ed.), *Natural disaster management* (pp. 82–98). London: Tudor Rose.

Dasgupta, S., Hettige, H., & Wheeler, D. (2000). What improves environmental performance? Evidence from the Mexican Industry. *Journal of Environmental Economics and Management, 39*(1), 39–66.

Dasgupta, S., & Wheeler, D. (2001). Small plants, industrial pollution and poverty. In Hillary (Ed.), *Small and Medium-sized firms and the environment* (pp. 289–304). Sheffield: Greenleaf Publishing.

Deegan, C. (2002). The legitimizing effect of social and environmental disclosures – A theoretical foundation. *Accounting, Auditing & Accountability Journal, 15*, 282–311.

Dilley, M., Chen, S., Deichmann, U., & Lerner-Lam, L. (2005). *Natural disaster hotspots: A global risk analysis*. Washington, DC: World Bank.

Downing, T. E, Butterfield, R., Cohen, S., Huq, S., Moss, R., Rahman, A., et al. (2001). *Vulnerability indices. Climate change impacts and adaptation*. UNEP Policy Series. Nairobi: UNEP.

Dryzek, J. S. (1997). *The politics of the earth*. Oxford: Oxford University Press.

Du Toit, A. (2004). Social exclusion' discourse and chronic poverty: A South African case study. *Development and Change, 35*(5), 987–1010.

Eakin, E., & Lemos, M. C. (2006). Adaptation and the state: Latin America and the challenge of capacity-building under globalization. *Global environmental change, 16*(1), 7–18.

ECLAC. (2010). *Statistical yearbook for Latin America and the Caribbean*, Economic Commission for Latin America and the Caribbean.

Edelstein, M. (1987). *Contaminated communities. The social and psychological impacts of residential toxic exposure*. Boulder, CO: Westview Press.

Erickson, K. (1994). *A new species of trouble. Explorations in disaster, trauma, and community*. New York: W. W. Norton.

Eskeland, G. S., & Jimenez, E. (1992). Policy instruments for pollution control in developing countries. *The World Bank Research Observer, 7*(2), 145–169.

Filgueira, C., & Peri, A. (2004). *América Latina: los rostros de la pobreza y sus causas determinants*. New York: United Nations Publications.

Funtowicz, S., & Ravetz, J. (1993). *Epistemología Política, ciencia con la gente*. Colección Fundamentos Ciencias del Hombre, Buenos Aires: CEAL.

Gallopin, G. (2006). Linkages between vulnerability, resilience, and adaptive capacity. *Global Environmental Change, 56*, 293–303.

Giddens, A. (1990). *The consequences of modernity*. Cambridge: Polity.

Giddens, A. (1991). *Modernity and self-identity self and society in the late modern age*. Cambridge: Polity Press.

Giddens, A. (1992). *The transformation of intimacy*. Cambridge: Polity Press.

Grynspan, R., & Kliksberg, B. (2008). Corporate social responsibility in Latin America. Not a waste of time or money. *Foreign Policy, July/August*, 167–169.

Guidi, M. (2008). Rethinking corporate social responsibility. A case study in Argentina from the point of view of the Civil Society, *Nómadas. Revista Crítica de Ciencias Sociales y Jurídicas, 19*(3) Downloadable at http://redalyc.uaemex.mx/redalyc/pdf/181/18101919.pdf

Hart, S. L. (1995). A natural-resource-based view of the firm. *Academy of Management Review, 20*, 986–1014.

Haslam, P. A. (2004). *The corporate social responsibility system in Latin America and the Caribbean*. FOCAL Policy Papers, FPP-04-1. March. Downloadable at http://www.focal.ca/pdf/csr_04.pdf

Hasnas, J. (1998). The normative theories of business ethics: A guide to the perplex. *Business Ethics Quarterly, 8*, 19–42.

Heyes, A. (1998). Making things stick: Enforcement and compliance. *Oxford Review of Economic Policy, 14*(4), 50–63.

Hochstetler, K. (2002). After the boomerang. Environmental Movement and Politics in the La Plata River Basin. *Global Environment Politics, 2*, 35–58.

Jagnicke, M (1985) *Preventive environmental policy as ecological modernisation and structural policy*. Berlin: WZB.

Leach, M., Bloom, G., Ely, A., Nightingale, P., Scoones, I., Sha, E., & Smith, A. (2007). Understanding Governance: pathways to sustainability, *STEPS Working Paper 2*. Brighton: STEPS Centre.

López Cerezo, J. A., & Luján-Lopez, J. L. (2000). *Ciencia y política del riesgo*. Madrid: Alianza.

Marris, P. (1996). *The politics of uncertainty. Attachment in private and public life*. London/New York: Routledge.

Martínez-Alier, J. (2002). *The environmentalism of the Poor. A study of ecological conflicts and valuation*. Cheltenham: Edward Elgar Publishing.

Minujín, A. (1999). ¿La gran exclusión? Vulnerabilidad y exclusión en América Latina. En D. Filmus (Ed.), *Los noventa. Política, sociedad y cultura en América Latina* (pp. 53–77). Buenos Aires: FLACSO/EUDEBA.

Morduch, J. (1994). Poverty and vulnerability. *American Economic Review, 84*, 221–225.

Natenzon, C. E. (2003). Inundaciones catastróficas, vulnerabilidad social y adaptaciones en un caso argentino actual. Cambio climático, elevación del nivel medio del mar y sus implicancias. *Climate Change Impacts and Integrated Assessment EMF (Energy Modeling Forum) Workshop IX*. Stanford University, Snowmass, CO, July 28–August 7, 16.

Newell, P., & Muro, A. (2006). Corporate social and environmental responsibility in Argentina. The evolution of an Agenda. *Journal of Corporate Citizenship, 24*, 49–68.

Peinado-Vara, E. (2006). Corporate social responsibility in Latin America. *Journal of Corporate Citizenship, 21*, 61–69.

Perrow, C. (1984). *Normal accidents. Living with high-risk technologies*. New York: Basic Books.

Porto de Souza, M. F. (2007). *Uma ecologia Politica dos Riscos*. Rio de Janeiro: Editora Fiocruz.

Porto de Souza, M. F., & Freitas, C. M. (2003). Vulnerability and Industrial Hazards in Industrializing Countries. An Integrative Approach. *Futures, 35*(7), 717–736.

Pratt L., & Fintel, E. (2002). Environmental management as an indicator of business responsibility in Central America. In P. Utting (Ed.), *The greening of business in developing countries. Rhetoric, reality and prospects* (pp. 41–57). London: Zed Books in association with UNRISD.

Prieto-Carron, M., Lund-Thomsen, B., Chan, A., Muro, A., & Bhushan, C. (2006). Critical perspectives on CSR and development: What we know, what we don't know, and what we need to know. *International Affairs, 82*, 977–987.

Pumpim de Oliveira, J. A., & Gardetti, M. A. (2006). Analysing changes to prioritise corporate citizenship. *The Journal of Corporate Citizenship, 21*, 71–83.

Quinn, J. J. (1997). Personal ethics and business ethics: The ethical attitudes of owner/managers of small business. *Journal of Business Ethics, 16*(2), 119–127.

Renn, O. (1992). Concepts of risk. A classification. In S. Krimsky & D. Golding (Eds.), *Social theories of risk* (pp. 53–79). Westport, CT: Praeger Publishers.

Ruiz-Tagle, M. T. (2003). *New approaches to environmental regulation in less developed countries. The case of Chile*, PhD. Dissertation, University of Cambridge.

Scherer, A. G., & Palazzo, G. (2007). Towards a political conception of corporate responsibility: Business and society seen from a Habermasian perspective. *Academy of Management Review, 32*(4), 1096–1120.

Schmidheiny, S. (2006). A view of corporate citizenship in Latin America. *Journal of Corporate Citizenship, 21*, 21–24.

Scoones, I., Leach, M., Smith, A., Stagl, S., Stirling, A., & Thopson, J. (2007). Dynamics systems and the challenge of sustainability, *STEPS Working Paper 1*. Brighton: STEPS Centre.

Scott, L. (2006). *Chronic poverty and the environment. A vulnerability perspective*, Chronic Poverty Research Centre, Working Paper 62.

Siegel, D. (2009). Green management matters only if it Yoelds more green: An economic/strategic perspective. *The Academy of Management Perspectives, 23*(3), 5–17.

Siegel, P. B., & Alwang, J. (1999). *An asset based approach to social risk Management: A conceptual framework*: Discussion Series No 9926 Social Protection. Washington, DC: World Bank.

Torres-Baumgarten, G., & Yucetepe, V. (2009). Multinational firms leadership role in corporate social responsibility in Latin America. *Journal of Business Ethics, 85*, 217–224.

UN/ISDR. (2004). *United Nations/international strategy for disaster reduction, living with risk: A global review of disaster reduction initiatives.* Geneva: United Nations.

Vazquez-Brust, D. A., & Liston-Heyes, C. (2008). Corporate discourse and environmental performance in Argentina. *Business Strategy and the Environment, 17*, 179–193.

Vazquez-Brust, D. A., Liston-Heyes, C., Plaza-Úbeda, J., & Burgos-Jiménez, J. (2010). CSR, stakeholders' management and stakeholders integration in Latin-America. *Journal of Business Ethics, 91*(2), 171–192.

Vazquez-Brust, D. A., Plaza-Úbeda, J. A., Natenzon, C. E., & Burgos-Jiménez, J. (2009). The challenges of businesses intervention in areas with high poverty and environmental deterioration: Promoting an integrated stakeholders approach in management education. In C. Wankel & J. Stoner (Eds.), *Management education for global sustainability* (pp. 175–206). New York: Information Age Publishing.

Vives, A. (2006). Social and environmental responsibility in small and medium enterprises in Latin America. *Journal of Corporate Citizenship, 21*, 39–50.

Wheeler, S. (2004). *Planning for sustainability.* New York and London: Routledge.

Wisner, B., Blaikie, P., Cannon, T., & Davis, I. (2004). *At risk: Natural hazards people's vulnerability and disasters.* London: Routledge.

Wynne, B. (1992). Uncertainty and environmental learning – Reconceiving science and policy in the preventive paradigm. *Global Environmental Change, 2*, 111–127.

Yapa, L. (2002), How the discipline of geography exacerbates poverty in the third world. *Futures. The Journal of Forecasting and Planning, 34*, 33–46.

Chapter 3
Statistical Information for the Analysis of Social Vulnerability in Latin America – Comparison with Spain

Anabel Calvo, Mariana L. Caspani, Julieta Barrenechea, and Claudia E. Natenzon

Abstract This chapter shows the results obtained in the search for empirical information on social aspects in Latin America relating to social vulnerability on a national scale. It also puts forward the reasoning behind this methodological approach, the procedures employed in the search and the problems encountered, establishing where possible a comparison with the Spanish case. It shows the results obtained in the selection of indicators and in the data corresponding to each of them for the vast majority of Latin American countries, and it compares the available indicators in Spain and Argentina covering the dimensions and the representative variables of different conditions of social vulnerability.

Keywords Social vulnerability · Poverty · Indicators · Latin-America · Spain

1 Introduction

Latin America is the region of the world with the greatest inequality, where the richest sector of the population enjoys the highest share of national income and the poorest sector the lowest share of wealth. Though Latin America has always been characterised by generalised poverty, high numbers of indigents, profound inequality and a tendency to social exclusion, since the 1990s these problems have taken on an importance that was unknown in previous decades. The crisis of formal employment, the emergence of structural unemployment and the persistence of an informal economy of poverty are some of the pillars of this society of inequality.

A. Calvo (✉)
Faculty of Philosophy and Letters, Institute of Geography "Romualdo Ardissone", University of Buenos Aires, Buenos Aires 1406, Argentina
e-mail: belcalvodiaz@gmail.com

M.L. Caspani (✉)
Department of Geography, Faculty of Philosophy and Letters, University of Buenos Aires, Buenos Aires 1406, Argentina
e-mail: mcaspani@gmail.com

The highest incidence of poverty is to be found in composite and extended families becoming even worse in single-parent families, in particular those where the head of family is female. From the point of view of geographical distribution, the most recent statistics (ECLAC, 2010) show that poverty and indigence are still more prevalent in rural areas of this region than in urban ones. The economic and social policies predominantly implemented in Latin America have been unable to respond effectively to the scenario of poverty in which wide sectors of the population are immersed.

Within this general common framework of Latin American countries, the social, economic and cultural characteristics of the different social groups present a great heterogeneity of situations. Theoretical discussions about the complex problem of poverty and exclusion, had historically ignored such heterogeneity or attempted to take it into consideration but failed due to conceptual limitations (Filgueira, 2006). Consequently, new discussions about innovative analytical frames, which allowed the development of new theories, concepts and methodological approaches, were introduced.

Filgueira (2006) makes a historical review of different approaches to studying social groups exposed to extreme deprivation. Throughout his analysis, he highlights the growing conceptual complexity of the issue. Filgueira identifies a shift from use of systems of statistics (such as sums of variables or social attributes) in the 1950s, to use of systems of indicators in the 1960s, which aimed at integrating different dimensions of the social problem and their standardisation. In the 1980s, the use of indices and indicators aimed to reflect a more structural situation but still were not able to clearly reflect the heterogeneity of the situations in need of description. The most frequently used measurements were poverty line[1] and unsatisfied basic needs.[2] In the case of poverty line, the measurement focuses on economic aspects related to the access to goods and services, identifying a dichotomy between the poor and non-poor society. In the second case, the selected social indicators allow to identify a variety of structures of shortcomings in the families within the UBN qualification, in terms of presence or otherwise of the defined characteristics.

In the 1990s, a last generation of indicators was developed to reflect a more dynamic perspective, which embraced heterogeneity and social complexity. The discussion on social vulnerability must be considered within this context, which underlies the proposal to use this notion in the analysis of social problems.

---

[1] The Poverty Line (PL) represents the level of income required by a household to meet its member's basic needs. It uses the average estimated cost of a basket of staple foods multiplied by the Engel coefficient (ratio of food expenditures to total expenditures). The resulting value would indicate the income necessary to cover a wide range of basic needs: food, housing, clothes, education, health, transport and leisure that together constitute the basic total family basket (CBT). The PL is related to the identification of "new poor", i.e. those sectors (mainly middle class) that have become poor due to the permanent loss of capital as a result of the recent process of economic adjustment.

[2] Unsatisfied Basic Needs (UBN) this measurement uses data from the census and the Permanent Household Survey (Encuesta Permanente de Hogares). The UBN index combines 5 indicators for each family: overcrowding, sanitary conditions of the household, schooling, employment and educational level of the head of family. It is used to identify the "structural poor", i.e. those who have always been poor.

According to Minujín (1998, 1999), social vulnerability allows us to analyse the dynamic complexity of situations of poverty, particularly those which arise from neoliberal reform programmes and structural adjustments that were strictly applied in the 1990s. It also allows us to identify an area of significant gradients (or intermediate situations) between the extremes "inclusion/exclusion" or "wealth/poverty".

According to Filgueira and Peri (2004:22) "the vulnerability concept makes a contribution to the analysis of social inequality avoiding the dichotomy poor-no poor". It integrates different aspects of social, economic, cultural and political reality, which are shown in poverty, exclusion and lack of social cohesion; and proposes the idea of vulnerable configurations capable of descending social mobility.

Social vulnerability is related to social exclusion and marginalisation, the place of articulation between assets and the structure of opportunities.[3] This means that the vulnerability of a household depends on its material and symbolic resources related to the structure of opportunities to which it has access.

In the field of studies on catastrophes other approaches can be found. For instance, according to Blaikie et al. (1998) social vulnerability is a set of previous characteristics belonging to a person or group of persons which determine their capacity to anticipate, survive, resist and recover from the impact of a given danger. It is a relative and specific term that always implies a certain vulnerability to a specific threat. To measure vulnerability in a situation of catastrophe, Downing et al. (2001) points out that specific indices can be drawn up for each particular case studied. The selection of indicators to make up an index of this type depends on how pertinent they are to fulfil the proposed aims and to reveal the aspects that constitute the vulnerability of the society in question.

One tool developed in this line of work is the SVI, or Social Vulnerability Index (Barrenechea et al. 2003; Natenzon et al. 2005), which was designed to assist in the analysis of industrial risk as part of the AECI project. This is a quantitative statistical assessment that allows us to identify the geographical distribution of different degrees of social vulnerability in a given set of administrative units by means of indicators that are chosen for this purpose. The administrative units that are identified as having the highest degree of social vulnerability can be taken as case studies to identify in greater depth, using qualitative techniques, what this social vulnerability consists of and how it has come into being.

The scope and limitations of an index of this kind are inherent to its construction, and will depend on:

1. the availability of equivalent information for each and every one of the administrative units that make up the area of study,
2. the criteria for selection of indicators; and
3. the internal ranges established for each of the indicators.

---

[3] Assets refer to the possession by individuals of material and symbolic resources that allow them to take part in society. The structure of opportunities is determined by the State, the market and society; it is not regulated by the individual (Filgueira, 2006:27).

It is useful as it provides a first approximation to the heterogeneity in the geographical distribution of social vulnerability, helping to prioritise or select the samples or case studies for which the analysis should be carried out in greater depth.

The indicators selected arise from the relationship between the representativeness of social, demographic and basic economic aspects on the one hand and the public availability of data on the other. The latter is one of the premises established from the outset in the first works that used this technique, with a view to making it accessible to local governments and civil organisations with few economic resources.

There follows an overview of the availability of statistical information expressing features of social vulnerability in Latin American countries, showing how this information was acquired and how it was selected to be used as an indicator.

2 Availability of Regional-Level Statistical Information; Comparison with Spain

The statistical information used to elaborate indicators of social vulnerability in Latin America was compiled from international institutions; it was analysed and selected according to its relevance in reflecting critical aspects of current social problems.

A basic assumption of this work was to consider that the definition and analysis of a system of social indicators imply a high degree of subjectivity, and they respond to conceptual frameworks that may be more or less explicit. In particular, regarding the use of indicators, we have followed the line of Gutiérrez-Espeleta, who considers that indicators are necessary tools of support, not only to explain the conditions or the situation of a society, but also to understand why those conditions exist" (2002:132).

Over the course of the last decade United Nations World Conferences have focused on different dimensions of human poverty, establishing eight[4] objectives of international development aimed at achieving the reduction of poverty between 2000 and 2015, mobilising international support. To fulfil these objectives, one of the strategies proposed has been to support programmes that create statistical capacity in developing countries and that elaborate several sets of indicators. Establishing a common system of assessment for these countries allows them to reduce the amount of specialised indicators, thus making it easier to monitor the progress achieved in meeting the objectives that have been set. These lists of indicators are used by the United Nations Assistance Framework for Development, governments, the UNO and other associated entities when selecting indicators to supervise

---

[4] To carry out these objectives it was suggested to develop actions that can be monitored through a set of indicators. These goals are proposed to eradicate Extreme poverty and hunger, achieve universal primary education, promote gender equality and empower women, reduce child mortality, improve maternal health; combat HIV/AIDS, malaria and other diseases; ensure environmental sustainability, develop a global partnership for development (http://www.undp.org/).

the development strategies in these countries, referring to variables such as welfare, work, education, health, gender, housing and basic services, population and economy.

Among the above-mentioned indicators are the Minimum National Social Data Set (MNSDS) of the United Nations and the basic set of indicators of the progress in development used by the Development Assistance Committee (DAC) of the Organisation for Economic Cooperation and Development (OECD), the World Bank and the UN.

Of all these statistical proposals, the present work has taken as reference the social indicators that make up the MNSDS, complemented by the Millennium Development Goals.[5] It also takes into account a series of additional indicators proposed by ECLAC (2005a) and others which feature in the Database of Basic Indicators of the PAHO. A further reference that has been considered is another work by ECLAC (2005b) which offers a panorama of indicators for Latin America, focusing on those areas in which there are unsatisfied needs and which require public social policies, rather than on areas of economic production or environmental policies.

On the other hand, Spain, as a member of the European Union, takes part in initiatives of international cooperation to foment the production of harmonised statistics in developing countries; such is the case of the Partnership in Statistics for Development in the twenty-first Century (PARIS 21), which constitutes a key factor in carrying out the measures that were agreed upon in the UNO to have available reliable data, complementary indicators and MNSDS. Nevertheless, as far as harmonisation of public statistics is concerned, Spain participates in the European Statistics System (ESS) and in the European System of Central Banks (ESCB), through which European statistics are elaborated and published. The Community Statistical Programme 2008–2012 is currently in effect and it aims to ensure coherence and comparability of statistical information in the Community (CE, 2007).

As regards the set of harmonised statistical operations for EU member countries that are related to the indicators of social vulnerability selected for Latin America, we should highlight:

(a) The European Community Household Panel (ECHP) up to 2002, whose overall aim was to provide the European Commission with a tool of statistical observation and comparable, harmonised information on the following aspects of living standards and conditions and social cohesion:

1. Income and mobility due to income. Economic situation.
2. Poverty, privation, minimum standards of social protection and non-discrimination.
3. Employment, activity, permanent professional training and work migration.
4. Retirement, pensions and the socio-economic status of senior citizens.
5. Level of education and its effects on socio-economic conditions.

---

[5] See: http://www.undp.org/spanish/mdg/basics.shtml

(b) The Living Conditions Survey (LCS) replaced the ECHP in 2002 due to the need to update the statistical sources in line with new political demands and in order to improve the quality of the information, particularly regarding the deadlines for which the data should be available. The fundamental aim of the LCS is to provide a reference source on living standards, the conditions of the work market and social cohesion in relation to the information requirements of active policies of the EU in these fields and their effects on the population.

This statistical research is flexible regarding the sources employed. The European Union Statistics Office (Eurostat) strongly recommends the use of existing statistical sources on a national scale, whether they be sample surveys or based on administrative registers. In the case of Spain, as there is no data source that corresponds to these needs, we have opted for a survey which has been carried out since 2004. The variables of the study are the following: economic situation, poverty, privation, minimum standards of social protection and non-discrimination, employment, activity, permanent professional training, work migrations, retirement, pensions, socio-economic status of senior citizens, level of education and its effects on socio-economic conditions.

3 Selection of Indicators for Latin America and the Caribbean

The information included here has been selected bearing in mind the multi-dimensional nature of social problems. On the one hand it considers relevant topics raised in international forums mentioned above: socio-economic conditions, education, health, nutrition and the environment; and on the other, the effective availability when the search was carried out. The indicators selected as a result are outlined in Table 3.1.

The work consisted of identifying and selecting renowned international institutions that produce statistical information for Latin America related to social problems. Those selected were:

1. FTAA – Free Trade Area of the Americas
2. IDB – Inter-American Development Bank
3. WB – World Bank
4. ECLAC – Economic Commission for Latin America and the Caribbean
5. LACSS – Latin American Council of Social Sciences
6. IMF – International Monetary Fund
7. OECD – Organisation for Economic Cooperation and Development
8. OAS – Organization of American States
9. WTO – World Trade Organisation
10. WHO – World Health Organisation
11. UN – United Nations
12. PAHO – Pan-American Health Organisation
13. UNFPA – United Nations Population Fund

3 Statistical Information for the Analysis of Social Vulnerability in Latin America... 41

Table 3.1 List of selected indicators

Topics	Source	Indicators
Socio-economic	WHO	*Per capita Gross Domestic Product*
	ECLAC	*Urban unemployment rate*
	PAHO	*Unemployment rate*
	ECLAC	*Urban population in situation of poverty*
	PAHO	*Rate of dependence*
Education	ECLAC	*Illiteracy in the population aged 15–24*
	PAHO	*Rate of literacy*
	ECLAC	*Net rate of registration in primary education*
Housing	ECLAC	*Population with sustainable access to the best supplies of drinking water*
	ECLAC	*Population with access to improved sanitary services*
Health	ECLAC	*Public health spending as a percentage of GDP*
	WHO	*Total spending on health per capita*
	WHO	*Life expectancy*
	ECLAC	*Child death rate*
	ECLAC	*Death rate of children under 5*
	PAHO	*Annual proportion of deaths of children under 5 due to infectious intestinal diseases*
	PAHO	*Death rate of women in childbirth*
	WHO	*Incidence of tuberculosis*
	WHO	*Total fertility rate*
	PAHO	*Rate of use of contraceptive methods among women*
	PAHO	*Proportion of pregnant women attended by qualified personnel*
	PAHO	*Proportion of doctors*
	PAHO	*Proportion of hospital beds*
	PAHO	*Proportion of under-weight newly-born children*

ECLAC Economic Commission for Latin America and the Caribbean, *PAHO* Pan-American health organisation, *WHO* world health organisation

For each of these institutions, a process of four stages was set up, in order to achieve a global view of their objectives and priorities as well as a detailed review of their websites in terms of functionality and social statistical production.

The following activities were carried out:

1. Analysis of the conceptual and operative structure of the website
2. Identification of specific sections to obtain statistical data
3. Evaluation of available variables and indicators
4. Selection of indicators according to the following criteria:

   4.1. Thematic pertinence: Given that the existing sources of information are countless, and that this might disperse the research throughout infinite

approaches, it was very important to focus on measurements of social aspects first.
4.2. Comparability: this issue refers to several aspects related with other criteria such as coverage, considered territorial units, measurement units, period, actualisation, obtaining and calculating methodology, etc.
4.3. Reliability: selected organisations were chosen, among other reasons, because of their international tested credibility and detailed specifications on the information provided.
4.4. Effective availability of access to the data in terms of: free or charged access; functionality of data format; independent tables for each country or global tables for all the countries in the region, difficulty to reach the data due to the multiple pages and links within each website.
4.5. Wide coverage of countries: priority was given to those indicators with greatest coverage of Latin America.
4.6. Updated data: the most up to date data were preferred, but also taking into account the least possible variability of dates for the same indicator.
4.7. Current view: where temporal series of data were available, the comparative evolution of the indicator was not taken into account. A specific year was chosen for all the set (or several if the information was not available for all countries for the same year).
4.8. Not estimated or projected data: problems arose in some cases for which the data were so current that it seemed to have been estimated. In these cases, it was decided to work with non-estimated data, even if it meant going back one or more years.
4.9. Non combined indicators: it was decided not to use combined indices such as the Human Development Index (HDI) proposed by the United Nations, opting to work with simple indicators.
4.10. Not disaggregated data for specific aspects (gender, age group, urban-rural, etc.): it was decided to work with general indicators without making distinctions into the same variable. As exception we considered some cases in which data were only available for the urban scope.

In short, this research has allowed us to identify some significant problems that condition the possibility of elaborating indicators of social vulnerability on a regional scale. The lack of data, the bias of the topics that are priorities for the particular organism consulted and the difficulties of effective access to the information conditioned the definition of the selected indicators.

As a result of the tasks carried out, of the problems encountered in accessing comparable data, or of the political problems linked to international relations, several of the countries of Latin America, according to the list of countries defined by the UN, had to be omitted from the final choice of nations presented: Anguilla, Antigua and Barbuda, the Netherlands Antilles, Aruba, Barbados, Dominica, Grenada, Guadalupe, French Guyana, the Cayman Islands, the Falkland Islands, Turks and Caicos Islands, the British Virgin Islands, the US Virgin Islands, Martinique, Montserrat, Puerto Rico, Saint Kitts and Nevis, Saint Vincent and the Grenadines, and Santa Lucía.

Finally, the database was constructed selecting an initial set of indicators according to each organism, and then comparing all of them and drawing up a table of synthesis. In this table the formats of information were standardised and grouped around the key topics mentioned previously. The compiled data, organised according to the procedures mentioned above are presented in Table 3.2.

Given the experience of the work carried out, we believe that new lines can be set up for future research work on the analysis of social vulnerability in the region, by means of statistical indicators, in order to study these questions in greater depth and complement the present study

As for now, we propose the following:

1. Specific analysis and compilation of data from highly complex, relevant organisms such as those that make up the UNO.
2. Construction of an integrative indicator of social vulnerability.
3. Comparative historical analysis of some specific indicators, to analyse evolution and possible trends.
4. Analysis of "drill-down"[6] possibilities in administrative units beyond the national scale.
5. Search for similar indicators as those used for Argentina and Spain to construct the index of vulnerability on a regional level, which might allow comparison with and among other countries.

Below the cases of Spain and Argentina are provided as an example of comparison between nations.

4 Selection of Indicators for the Cases of Spain and Argentina

Once the set of harmonised indicators was identified for the national level of Latin American countries, we proceeded to select those that would make up the index of social vulnerability calculated for more disaggregate geographical levels within the countries. The reference methodology used was PIRNA, as mentioned previously. We applied this methodology as a template to select the best available indicators of social vulnerability using free, public statistical sources in Argentina, and later to identify the corresponding indicators in Spain.

In the case of Argentina, the Instituto Nacional de Estadísticas y Censos (INDEC) is a public organism that depends on the Ministry of Economy and Production, and it is responsible for coordinating the National Statistical System (SEN) which is made up of the statistical services of national, provincial and municipal organisms. The INDEC is directly responsible for the methodological and regulatory development

---

[6] By means of these processes and procedures the research goes from summarised/synthetic information to more detailed information, to focus on a particular aspect or on a smaller unit of information.

Table 3.2 Results of the compilation and selection of indicators for Latin America and the Caribbean

Topic	Socio-economic					
Source	WHO	ECLAC	PAHO		ECLAC	ECLAC
Year	2006					2008
Indicator	Per capita Gross Domestic Product	Rate of urban unemployment	Year	Proportion of unemployed	Year	Rate of dependence (estimated)
Unit	US$	%		%		Dependent population for each 100 active persons
Countries						
Argentina	11,670	8.5	2007	10.2	2006	56.2
Bahamas	n.d.a.	7.9	2007	*		*
Belize	7,080	8.5	2007	11	2005	67.2
Bolivia	3,810	7.7	2007	5.4	2002	70.7
Brazil	8,700	9.3	2007	8.9	2004	50.9
Chile	11,300	7.1	2007	6.9	2005	47.3
Colombia	6,130	11.4	2007	9.5	2005	52.1
Costa Rica	9,220	4.8	2007	6.6	2005	49.2
Cuba	n.d.a.	1.8	2007	1.9	2004	42.9
Dominican Rep.	5,550	15.6	2007	17.9	2005	62.6
Ecuador	6,810	7.4	2007	7.7	2005	60.7
El Salvador	5,610	5.7	2006	6.6	2006	62.8
Guatemala	5,120	4.4	2004	3.4	2003	87.1
Guyana	3,410	n.d.a.	2007	*		*
Haiti	1,070	n.d.a.	2007	n.d.a.		n.d.a.
Honduras	3,420	4.11	2007	4.2	2005	73.8
Jamaica	7,050	9.9	2007	*		*
Mexico	11,990	4.8	2007	3.5	2005	54.8
Nicaragua	2,720	6.9	2007	8	2003	67.1
Panama	8,690	7.8	2007	10.3	2005	56.1
Paraguay	4,040	7.2	2007	7.9	2003	65.2
Peru	6,490	8.4	2007	11.4	2005	56.3
Surinam	7,720	12	2007	*		*
Trinidad and Tobago	16,800	5.6	2007	*		*
Uruguay	9,940	9.6	2007	12.2	2005	58.2
Venezuela	10,970	8.4	2007	15	2004	55.2

* Countries not contemplated by the organism providing the statistical information, *n.d.a.* no data available

3 Statistical Information for the Analysis of Social Vulnerability in Latin America...

Table 3.2 (continued)

Topic	Education				
Source	ECLAC	PAHO		ECLAC	
Year	2005				
Indicator	Rate of illiteracy among the population aged 15–24	Rate of literacy	Year	Net rate of registration for primary education	Year
Unit	%	%		%	
Countries					
Argentina	1.3	97.6	2007	98.5	2005
Bahamas	2.5	*		87.9	2006
Belize	1.4	n.d.a.	2007	97	2006
Bolivia	3	90.3	2007	94.9	2006
Brazil	3.9	90.5	2007	94.4	2005
Chile	0.8	96.5	2007	n.d.a.	
Colombia	2.4	93.6	2007	88.5	2006
Costa Rica	1.4	95.9	2007	n.d.a.	
Cuba	0.2	99.8	2007	96.6	2006
Dominican Rep.	7.5	89.1	2007	77.5	2006
Ecuador	2.1	92.6	2007	96.8	2006
El Salvador	10	85.5	2007	94	2006
Guatemala	18.4	73.2	2007	94.4	2006
Guyana	0.2	*		n.d.a.	
Haiti	31.2	n.d.a.		n.d.a.	
Honduras	12.7	83.1	2007	96.4	2006
Jamaica	4.9	*		90.3	2005
Mexico	2.3	92.4	2007	97.7	2006
Nicaragua	26.8	80.5	2007	89.8	2006
Panama	2.6	93.4	2007	98.5	2006
Paraguay	2.4	93.7	2007	94.3	2005
Peru	2.4	90.5	2007	96.3	2006
Surinam	n.d.a.	*		96.4	2006
Trinidad and Tobago	0.2	*		84.6	2005
Uruguay	0.8	98	2007	99.7	2006
Venezuela	1.4	93	2001	91.1	2006

* Countries not contemplated by the organism providing the statistical information, *n.d.a.* no data available

Table 3.2 (continued)

Topic	Housing			
Source	ECLAC	Source PAHO-Year	ECLAC	Source PAHO-Year
Year	2006		2006	
Indicator	Population with sustainable access to better supply of drinking water in urban and rural areas		Population with access to improved sanitary services in urban and rural areas	
Unit	%		%	
Countries				
Argentina	96		91	
Bahamas	n.d.a.		100	
Belize	91	2004	47	2004
Bolivia	86		43	
Brazil	91		77	
Chile	95		94	
Colombia	93		78	
Costa Rica	98		96	
Cuba	91		98	
Dominican Rep.	95		79	
Ecuador	95		84	
El Salvador	84		86	
Guatemala	96		84	
Guyana	93		81	
Haiti	58		19	
Honduras	84		66	
Jamaica	93		83	
Mexico	95		81	
Nicaragua	79		48	
Panama	92		74	
Paraguay	77		70	
Peru	84		72	
Surinam	92		82	
Trinidad and Tobago	94		92	
Uruguay	100		100	
Venezuela	83	2004	68	2004

* Countries not contemplated by the organism providing the statistical information, *n.d.a.* no data available

Table 3.2 (continued)

Topic	Health (1/3)						
Source	ECLAC	WHO	WHO	ECLAC	ECLAC	PAHO	PAHO
Year Indicator	2006 Public health spending as % of GDP at current prices	2005 Total health spending per capita	2006 Life expectancy	2006 Child death rate	2006 Death rate of children under 5	Annual proportion of deaths of children under 5 due to infectious intestinal diseases	Year
Unit	%	US$	Years	‰	‰	%	
Countries							
Argentina	4.58	1,529	75	14	16	1.2	2005
Bahamas	n.d.a.	1,404	74	13	14	*	
Belize	2.7	377	69	14	16	3.7	2001
Bolivia	4.2[a]	206	66	50	61	7.1	2003
Brazil	3.6[a]	755	72	19	20	4.4	2005
Chile	2.8	668	78	8	9	0.5	2005
Colombia	3.38	581	74	17	21	4.3	2005
Costa Rica	6	684	78	11	12	1.1	2007
Cuba	7.10	333	78	5	7	1.4	2006
Dominican Rep.	1.93	378	70	25	29	4.7	2004
Ecuador	n.d.a.	274	73	21	24	6	2005
El Salvador	4.1	364	71	22	25	8.8	2006
Guatemala	2.2[a]	244	68	31	41	14.7	2004
Guyana	2.7	238	64	46	62	*	
Haiti	n.d.a.	71	61	60	80	n.d.a.	1990
Honduras	3.9[a]	226	70	23	27	29.3	
Jamaica	n.d.a.	210	72	26	31	*	
Mexico	2.9	725	74	29	35	4.1	2006
Nicaragua	3.4	240	71	29	36	8	2005
Panama	3.2	660	76	18	23	6.5	2004
Paraguay	3	334	75	19	22	4.9	2006
Peru	1.1	284	73	21	25	3.3	2004
Surinam	1.47	325	68	29	39	*	
Trinidad and Tobago	1.1	763	69	33	38	*	
Uruguay	3.8	885	75	11	12	1.7	2004
Venezuela	4.4[a]	325	74	18	21	8.5	2005

* Countries not contemplated by the organism providing the statistical information, *n.d.a.* no data available
[a] PAHO

Table 3.2 (continued)

Topic	Health (2/3)					
Source	PAHO	PAHO	WHO	WHO	PAHO	PAHO
Year Indicator	Death rate of women in childbirth	Year	2006 Incidence of tuberculosis (number of possible new cases)	2006 Total fertility rate	Rate of use of contraceptive methods among women	Year
Unit	per 100,000 newborn		per 100,000 inhabitants	children/woman	%	
Countries						
Argentina	7.8	2006	39	2.3	67.8	2004
Bahamas	*		38	2	*	
Belize	n.d.a.	2007	49	3	56	2004
Bolivia	229	2003	198	3.6	21.3	2006
Brazil	74.7	2005	50	2.3	81	2006
Chile	18.1	2006	15	1.9	63.9	2006
Colombia	73.1	2005	45	2.3	56	2005
Costa Rica	n.d.a.	2007	14	2.1	96	2006
Cuba	30.2	2007	9	1.5	77.1	2005
Dominican Rep.	72.8	2007	89	2.9	51.2	2002
Ecuador	73	2006	128	2.6	66	2003
El Salvador	71.2	2005	50	2.7	67.3	2003
Guatemala	148.8	2005	79	4.3	40.1	2002
Guyana	*		164	2.4	*	
Haiti	n.d.a.		299	3.7	n.d.a.	
Honduras	108	1997	76	3.4	43.2	2005
Jamaica	*		7.3	2.5	*	
Mexico	58.6	2006	21	2.3	70.9	2006
Nicaragua	90.4	2006	58	2.8	72.4	2006
Panama	83.6	2006	45	2.6	48	2003
Paraguay	121.4	2006	71	3.2	72.8	2006
Peru	185	2000	162	2.5	70.5	2004
Surinam	*		64	2.5	*	
Trinidad and Tobago	*		8.4	1.6	*	
Uruguay	n.d.a.	2007	4.3	2.1	78	2005
Venezuela	59.9	2005	41	2.6	30	2005

* Countries not contemplated by the organism providing the statistical information, *n.d.a.* = no data available

Table 3.2 (continued)

Topic	Health (3/3)							
Source	PAHO	PAHO	PAHO	PAHO	PAHO	PAHO	PAHO	PAHO
Year								
Indicator	Proportion of pregnant women attended by qualified personnel	Year	Proportion of doctors	Year	Proportion of hospital beds	Year	Proportion of under-weight newly-born children	Year
Unit	%		Per 10,000 inhab.		Per 1,000 inhab.		%	
Countries								
Argentina	88.4	2005	32.1	2004	4.1	2000	7.2	2006
Bahamas	*		*		*		*	
Belize	99.2	2007	9.3	2006	1.2	2007	6.9	2007
Bolivia	79.1	2003	7.6	2001	1.1	2007	4.1	2004
Brazil	97.4	2005	16.4	2006	2.4	2005	8.1	2005
Chile	96	2006	9.3	2004	2.3	2006	5.9	2006
Colombia	93.5	2005	12.7	2003	1	2007	8.1	2005
Costa Rica	91.7	2006	20	2005	1.3	2006	6.8	2006
Cuba	100	2007	62.7	2005	4.9	2007	5.2	2007
Dominican Rep.	98.9	2007	20	2005	1	2007	10.8	2002
Ecuador	59	2006	15.4	2003	1.7	2003	5.4	2006
El Salvador	51.9	2007	12.6	2002	0.7	2007	8	2007
Guatemala	70.4	2007	9.7	2005	0.7	2007	4.9	2007
Guyana	*		*		*		*	
Haiti	n.d.a.		n.d.a.		n.d.a.		n.d.a.	
Honduras	91.7	2005	8.5	2006	1	2002	10	2005
Jamaica	*		*		*		*	
Mexico	94.2	2006	14	2006	1.6	2006	8.3	2006
Nicaragua	90.2	2006	16.4	2003	1	2006	8.3	2006
Panama	83.9	2007	13.8	2005	2.2	2005	9.3	2006
Paraguay	85.3	2007	6	2006	1.3	2007	6	2006
Peru	91	2006	11.5	2004	1.2	2007	8.4	2006
Surinam	*		*		*		*	
Trinidad and Tobago	*		*		*		*	
Uruguay	94.9	2006	38.7	2005	2.9	2007	8.3	2006
Venezuela	25.5	1997	20	2001	0.9	2003	8	2004

* Countries not contemplated by the organism providing the statistical information, *n.d.a.* no data available

of the production of official statistics, and for the organisation and management of the national operations of collection of data via censuses and surveys, the elaboration of basic social and economic indicators, as well as the production of other basic statistics. The production of statistical information is carried out by various methods (censuses, surveys, administrative registers, etc.), which allow indicators to be devised in relation to different topic areas. The most important ones on a national level are the National Census of Population and Housing, the National Economic Census and the National Agricultural Census. On their website dynamic databases are available for the National Census of Population and Housing as well as specific reports for other sources.

The National Population and Housing Census was carried out every 10 years for the last 50 years and, although more irregularly, since 1869. The last one dates from October 27th, 2011.

In the case of Spain, three types of population data can be distinguished that are included in the production of their National Statistics System:

– *statistical exploitation of the administrative data* of the Municipal Population Registry (Padrón Municipal de Habitantes).
– *statistical operations of synthesis,* to measure the present or future population using the best available information at any given moment in time. Depending on the time scale this information may be provided by Population Forecasts (Proyecciones de Población), Estimations of Current Population (Estimaciones de la Población Actual) and Inter-census Estimations of Population (Estimaciones Intercensales de Población).
– *Population Census* carried out every 10 years for the last one and a half centuries. The last one dates from November 1st, 2001.

The website of the National institute of statistics (INE) offers free access to an exhaustive database that is organised by topic areas, registering and providing details of the data source of data compiled by the INE and by other national and international bodies.

The specific analysis of each country can be seen in the corresponding chapters. Table 3.3 cites the indicators selected for the calculation of the index of social vulnerability in the countries of Latin America and Spain, comparing the name of the indicators as they appear in the Argentinean and Spanish public statistical sources.

The table shows that the differences between the census sources in Argentina and Spain are not very significant in relation to the selected indicators. Only in the case of the indicators of transitory dependent population is there a minor difference in the age group considered, as in Argentina it refers to the 0–14 age group, whereas in Spain it extends to the age of 16.

The most significant difference found relates to the indicator "health cover". In the case of Spain nobody lacks health cover as all citizens have access to the public health service. The indicators of the National Health Survey have been consulted, but this is a sample application which is representative on the national scale and on that of the autonomous communities selected; therefore it is not possible to count

Table 3.3 Dimensions, variables and indicators for Argentina and Spain

Dimension	Variable	Indicator in Argentina[a]	Indicator in Spain[b]
Demographic	1. Transitory dependent population	Population in households by age group and gender, according to the type of household and the relationship with the head of family	Population by age group – Difference in the age group 0–16
	2. Definitive dependent population	Population in households by age group and gender, according to the type of household and the relationship with the head of family	Population by age group – Difference in the age group 0–16
	3. Single-parent homes	Head of household of incomplete nucleus, by civil status, by gender and age group	Household with 1 man or woman 1 responsible for 1 or more minors
Economic capacity	4. Health cover	Population with health cover from a social charity and/or with a private or mutual health plan by gender and age group	No coverage is impossible as there is a public health system. "No access to the health service in the last 12 months" was chosen as an alternative indicator. Non-representative indicator[c]
	5. Literacy/education	Population of 10 or over by literacy and gender	Total number of illiterates
	6. Work/employment	Population of 14 or over in employment or not economically active, by gender and age group	Employment status of the head of family. Unemployed per household
Living standards	7. Overcrowding	Households with more than 3 persons per bedroom	Average surface area of household per inhabitant
	8. Supply of drinking water	Households with the presence of this service in the segment	Housing problems: number of households without running water
	9. Sewage services	Households with the presence of this service in the segment	Housing problems: number of households without sewage disposal

[a] These correspond to the name in the National Population Census, 2001
[b] Except for variable 4, these correspond to the name in the National Population Census, 2001
[c] National Health Survey, 2003 and 2006. The sample is representative on a national level and for the Autonomous Communities selected. Without a greater level of disaggregation

on data for the maximum levels of disaggregation which have been included in this study (province, municipality, census unit). In this way, for the comparison to be consistent we have opted to assign the same register to all the Spanish study units, indicating the existence of health cover for all of them. This difference implies a major advantage for Spain when comparing levels of social vulnerability with those of Latin American countries.

The information obtained on social vulnerability in each of the countries studied (considered here in absolute terms) has been related to the data of industrial hazardousness, obtaining values that allow us to establish different degrees of industrial risk, as is explained in greater detail in the respective chapters referring to Spain, Argentina and Bolivia.

References

Barrenechea, J., Gentile, E., González, S., & Natenzon, C. E. (2003). Una propuesta metodológica para el estudio de la vulnerabilidad social en el marco de la teoría social del riesgo. In S. Lago Martínez (Ed.), *En torno a las metodologías. Abordajes cualitativos y cuantitativos* (pp. 179–196). Buenos Aires: Proa XXI.

Blaikie, P., Cannon, T., Davis, I., & Wisner, B. (1998). *Vulnerabilidad. El entorno social, político y económico de los desastres*. Bogotá: LA RED/ITDG.

CE. (2007). DECISIÓN No 1578/2007/CE DEL PARLAMENTO EUROPEO Y DEL CONSEJO of 11-12-2007, programa estadistico comunitario 2008–2012, DO L 344/15. http://eur-lex.europa.eu/LexUriServ/LexUriServ.do?uri=OJ:L:2007:344:0015:0043:ES:PDF

CEPAL – Comisión Económica para América Latina y el Caribe. (2005a). *Objetivos de Desarrollo del Milenio. Una Mirada desde América Latina y el Caribe*. Santiago de Chile.

CEPAL – Comisión Económica para América Latina y el Caribe. (2005b). *Propuesta para un Compendio Latinoamericano de Indicadores Sociales. Unidad de Estadísticas Sociales*. Serie de Estudios Estadísticos y Prospectivos N° 41. Santiago de Chile.

Downing, T. E, Butterfield, R., Cohen, S., Huq, S., Moss, R., Rahman, A., et al. (2001). *Vulnerability indices. Climate change impacts and adaptation*. UNEP Policy Series. Nairobi: UNEP.

ECLAC. (2010). Statistical yearbook for Latin America and the Caribbean, Economic Commission for Latin America and the Caribbean.

Filgueira, C. (2006). Estructura de oportunidades y vulnerabilidad social. Aproximaciones conceptuales recientes. *Política y Gestión*: Homo Sapien, 9, 18–64.

Filgueira, C., & Peri, A. (2004). *América Latina: los rostros de la pobreza y sus causas determinants*. United Nations Publications.

Gutierrez-Ezpeleta, E. (2002). Indicadores sociales: una breve interpretación de su estado de desarrollo. In C. Sojo (Ed.), *Desarrollo social en América Latina: temas y desafíos para las políticas públicas* (pp. 107–148). San José de Costa Rica: FLACSO. http://168.96.200.17/ar/libros/costar/america/cap2.pdf

Minujín, A. (1998). Vulnerabilidad y exclusión en América Latina. In E. Bustello & A. Minujin (Eds.), *Todos entran. Propuesta para sociedades incluyentes* (pp. 161–205). Bogotá: LA RED/ITDG.

Minujín, A. (1999). ¿La gran exclusión? Vulnerabilidad y exclusión en América Latina. In D. Filmus (Ed.), *Los noventa. Política, sociedad y cultura en América Latina* (pp. 53–77). Buenos Aires: FLACSO/EUDEBA.

Natenzon, C. E, Marlenko, N., González, S. G, Ríos, D., Barrenechea, J., Murgida, A., et al. (2005). Vulnerabilidad social estructural. In V. Barros, A. Menéndez, & G. Nagy (Eds.), *El Cambio Climático en el Río de la Plata* (pp. 113–118). Buenos Aires: AIACC/CIMA.

Chapter 4
Evaluating the Firm's Environmental Hazardousness: Methodology

Sergio D. López and Diego A. Vazquez-Brust

Abstract After a brief introduction summarising the dominant approach to development of risk maps and their relationship to the conceptual approach used in this project, the chapter details the empirical procedure used for calculating industrial hazardousness maps. This methodology measures the sum of potential hazard in a given geographical area, using an algorithm to extend the influence of the potential hazard of each industry to the surrounding area, also overlapping the effects of various industries within an area of influence. This allows the location of areas of potential hazardousness due to the cumulative effects of small and medium-sized firms in each area that had not been identified by previous methodologies based only on the size or potential impact of individual companies.

Keywords Risk assessment · Industrial hazardousness · SMEs · GIS · Urban risks

1 Evaluating Industrially Generated Environmental Risk

According to the United States' EPA (Environmental Protection Agency), risk managers are individuals, teams, or organisations with responsibility for or authority to take action in response to an identified risk. This term is often used to designate decision makers in state agencies or those who have the legal authority to protect or administrate a resource. However, this definition could also cover those with the possibility to take action to mitigate risk, thus including representatives of national, regional or local governments, as well as representatives of commercial or industrial organisations, of NGOs, etc. (EPA 1998).

A great deal of research has been carried out on risk assessment, but it has tended to be from the rather limited perspective of the effects of specific stressors (generators/sources of risk) such as, for instance, analysis of the action/effects of certain contaminants, or the elaboration of models of both hydric and atmospheric pollution. Nonetheless, there has been little research aimed at elaborating a more universal tool to identify generic areas in which to carry out more exhaustive studies.

S.D. López (✉)
Unidad de Coordinación de Programas y Proyectos con Financiamiento Externo: Programa de Infraestructura Hídrica de las Provincias del Norte Grande, Ministerio de Planificación Federal, Inversión Pública y Servicios, Av. Roque Saenz Peña 938, Piso 6, CABA CP 1035
e-mail: sergiodlopez@yahoo.com

This shortcoming is also evident from the type of evaluation tools and assessment mechanisms included in the scope of environmental legislation in Latin America. In the specific Latin American cases covered in the present work (Bolivia and Argentina), we see that Bolivia has no regulatory tool to evaluate situations of potential risk, while in Argentina each province is responsible for elaborating its own risk assessment procedures. The province of Buenos Aires constitutes one of the more progressive cases. There, Provincial Law 11.459 on Industrial Establishments and Normative Decree 1741/96 establish precise mechanisms to assess the environmental complexity of each industry by determining a coefficient called Level of Environmental Complexity (LEC). This provides an idea of how complex an industry is and therefore of the degree to which it may have a negative impact on its environment. But what happens when an industrial agglomeration creates an overlapping of these negative effects? The legislation is not equipped with suitable tools to assess the accumulated impact created by the geographical proximity of stressors

The aim of this chapter, therefore, is to study and elaborate a tool that allows integral risk assessment from a spatial perspective, with a view to determining which areas are most likely to suffer some kind of negative impact as the result of the different industrial activities installed in a given geographical area.

Risk assessment from this spatial perspective, together with the tools available in GIS, has numerous advantages over the traditional approach of presenting results using tables and isolated values. It enables assessors and decision makers, those responsible for environmental risk control policies, to obtain results by simply consulting a map, allowing better allocation of human and economic resources.

A Geographical Information System is used as an integrating tool, by means of which, along with geo-statistical techniques, "risk surfaces" are generated. Once the surfaces have been calculated for each impact, they are combined in a single map and weighted according to their importance or incidence in the value of overall risk.

The methodology adopted for each case study (Spain, Bolivia, Argentina) will be as thorough as the level of available information permits, and it should be understood as a proposal. It should be noted that the use of GIS to assess the spatial component of risk involves using making certain assumptions and simplifications. For instance, the gradients of contamination present anisotropies due to the influence of directional factors such as prevailing winds or the run-off direction of underground water. In other cases the available information is not sufficiently disaggregate. The people who are responsible for administering and controlling industrial activity and health in general must, therefore, anticipate such scenarios and take preventive and/or mitigating measures, and to do so risk assessment methodologies need to be implemented.

1.1 What Is Risk Assessment?

According to the Environmental Protection Agency, risk assessment is a process that analyses the likelihood of adverse effects as the result of exposure to one or more sources of potential environmental impact (EPA 1998).

4 Evaluating the Firm's Environmental Hazardousness: Methodology

The assessment of these risks may vary from qualitative judgements to the quantitative calculation of probability. Although risk assessment may include quantitative estimators of risk, quantification of all risks is not always possible. However, it is always preferable to reach conclusions based either on qualitative judgements or in imperfect quantitative proxies (with associated uncertainties) than to ignore these risks that cannot be easily understood or quantitatively estimated

As explained in detail in Chapter 2, risk can be considered as being made up of 5 components:

Hazardousness or perilousness **refers** to the state of being dangerous' inherent in the phenomenon; *Vulnerability* is defined as the differentiated capacity to face up to the impact phenomenon; *Exposure to risk* refers to the spatial and temporal distribution of what may potentially be affected. These three factors constitute *Evaluated (potential) risk. Evaluated (potential) risk does not necessarily imply real/managed risk*. Two additional components cover the breach between potential and real or managed risk: *uncertainty* refers to the limitations of the scientific method to interpret and assess the phenomenon studied; and *governability* refers to the existence of social and institutional structures that lessen the likelihood of damage occurring.

The assessment of risks derived from natural hazards acknowledges that in many cases it is not possible to modify the hazard in order to reduce the risk (e.g. a volcano eruption). In such cases of unmanageable natural hazard there is nothing left to do except modify the conditions of vulnerability of the exposed elements (e.g. improve evacuation and early alarm systems). On the other hand, in the case of technological threats such as industrial pollution, human intervention can – in theory – more easily reduce the level of hazard generated by the technological threat than the vulnerability of exposed communities. For that reason, pollution risk assessment has traditionally focused on the evaluation of levels of hazard and the design of instruments – regulation, technology – to reduce it. However, social and institutional structures can create powerful disincentives to reduce the hazard of pollution. For instance, more stringent regulation to abate pollution could be resisted by local communities who think pollution is a cost to be paid for development and are afraid of firms closing down under regulatory pressure or because politicians are not certain about the real impacts of pollution. Therefore, a thorough assessment of industrial risk requires understanding of vulnerability of the exposed elements, but also awareness of conditions of governability. When governability is low and uncertainty high, there are strong disincentives to modify the likelihood of hazard and also to reduce the vulnerability of communities. Thus, reducing exposure of vulnerable populations is often the only way left to reduce risk.

One of the best ways to assess *Exposure* to risk and to summarise different types of information that affect a given area is by using a map. As far as risk assessment is concerned, this means using so-called "risk cartography" in which the indicators of vulnerability and hazardousness are spatially overlapped (in actual fact the present work refers to "evaluated (potential) risk cartography", but in the review that follows we use the terminology used by the cited authors).

Generally speaking, "risk cartography aims to identify geographical areas that are susceptible to suffering damage should a threat become a reality" (Lowry et al. quoted by Bosque Sendra et al. 2000). Thus, risk cartography implies identifying and locating the three components of the problem mentioned above and determining their spatial characteristics.

Although a plethora of research works have employed risk cartography (e.g. Lirer and Vitelli 1998; Bankoff et al. 2004; Aceves-Quesada 2006; Meyer et al. 2009) most have been concerned with natural disasters (floods, volcanoes, seismicity, etc.), while few have explored the cartography of technological risks (Malczewski 2006). In the case of natural risks the extent of the affected areas is clear, since the physical medium itself is the risk factor. On the other hand, the nature of the factors generating industrial or technological impact risks is so diverse that it becomes difficult to establish a zone of influence.

A stream of research has assessed the risks generated by industries in urban zones on both a national and an international level (e.g. Sengupta and Patil 1996; Christou and Mattarelli 2000; Fairhurst 2003; Cozzani et al. 2006; Basta et al. 2007). These works include common factors that can be applied to the present research. For instance, the idea of risk due to proximity to industrial activities, the influence of each risk factor limited to a certain area, the use of reference values established in environmental protection regulations and discrimination on qualitative levels of the final results.

The pioneering work of Bosque Sendra et al. (2000) assesses risk based on certain selected sources. Its effects spread over an area comprising the visual catchment area of each impact, and to this they add the overlapping effects to generate a final risk surface divided into quantitative categories (zero – low – medium – high risk).

Bosque Sendra et al. (2004) go a step further than in their previous work by incorporating into their definition of dangerous activities and their spatial location certain aspects that are contemplated in current Spanish and European legislation (RAMINP and Directive 96/82/CE).

A number of studies in the field of environmental justice[1] have mapped industrial risks in Latin-America (e.g. Ulberich 2000; Lara-Valencia et al. 2009 and references therein) not only from the viewpoint of possible stressors (sources/generators of risk), but also incorporating the vulnerability of the receptors. In Argentina, Ulberich (2000) analyses the incidence of industrial settlements on the urban environment of the city of Tandil (Buenos Aires, Argentina). This study uses parameters defined in the legislation as weighting tools and the influence of each industry is extended over an area that is determined by that industry's category.

---

[1] Environmental justice literature can be classified into two main streams: (a) Research focused on the pattern of environmental hazard distribution, in other words, whether vulnerable populations are disproportionately affected by environmental threats (e.g. US GAO 1983). (b) Analysis of temporal/spatial patterns of causality of environmental hazard, in other words whether vulnerable populations attract technological hazards or whether they follow industrial pollution in their search for jobs and access to infrastructure such as electricity and roads (e.g. Been and Gupta 1997).

The works mentioned above have certain factors in common with the present study and these have been analysed to elaborate the methodology. In addition our methodology seeks to address Bowen's (2002) diagnosis of weaknesses in empirical studies of environmental injustice. These include, inappropriate unit of spatial analysis (e.g., using a county as unit of mapping, does not take into account the place-specific nature of threats), spatial aggregation problems, incompletely specified models (aggregated effects are underestimated), lack of documentation, improperly conceptualised and selected comparison regions, unreliable data. Accordingly our methodology uses as unit of spatial analysis the minimal unit allowing data capture in a Geographical Information System (raster), varies the hazard measurement models according to the availability and reliability of data and allows the identification of hazardousness due not only to large industries but also to numerous small industries

2 Methodology of Analysis of Industrial Hazardousness/Perilousness

The perilousness of an industry is measured by calculating an indicator that represents the potential harmful impact of the industry's activity (IP). This indicator is based on the best available information for each case study and level of intervention according to the methodology described in the respective chapters.

The most basic spatial assessment tool is a Pin Map based on the industries addresses. This implies the geolocation of points on a base map according to the industry's address or Zip Code, and linking data (e.g. type of industry, size, power consumption, etc) to it. Therefore, this Pin Map of the industries, which reflects the potential environmental impact allocated to each industry, provides a first impression of industrial hazardousness. However, this only allows us to assess the degree of potential peril/hazard at the specific location of each industry. But what occurs in the spaces between those where industries are located? It is precisely here that many residential areas are located and they are susceptible to the impact of industrial activity.

In order to assess the aggregated environmental hazardousness in these areas, taking into account the concept of exposure, techniques of spatial analysis were applied. The idea was to develop a method that allows the influence of a potential hazard or peril generated by an industry at a given point (IP) to be extended to the neighbouring area, while at the same time overlapping the effects of several industries on a given area of influence in order identify the aggregated environmental hazard in a particular location (EH).

This type of application is also used in other fields, such as the analysis of crime-related events in order to detect areas of greater danger (or "hotspots") based on specific data of reported crimes (Williamson et al. 1999).

For the purposes of our application a "raster" data model was used. This is particularly suited to the representation of continuous values over a given surface, for example in studies of topography, concentrations, humidity, etc. This model divides the surface into discrete cells, each of which stores the corresponding value of the represented function. The Spatial Analyst extension of the ArcGIS 9 programme was used to carry out the spatial analysis. This extension includes several tools for the analysis and generation of continuous surfaces based on the interpolation of specific data. The following hypotheses were taken into account.

This extension includes several tools for the analysis and generation of continuous surfaces based on the interpolation of specific data. The following hypotheses were taken into account:

H1: The degree of hazard or peril generated by a particular industry, evaluated at a given point, is a function of the distance between the point where the hazard is evaluated and the industry that generates the hazard, i.e. the hazard ranges from a maximum value at the industrial site (IP), diminishing as the distance from the site increases, eventually reaching zero. Figure 4.1 shows the hypothesised variation of evaluated environmental hazard as a function of the distance to the polluting industry. The distance from the industry to the point where it has no potential harmful effects is called the industry's Radius of Influence (R).

H2: The Radius of influence of an industry's potential hazard or peril depends on the magnitude of the maximum potential hazard (IP: Evaluated Hazard at the point where the industry is located); the greater the maximum potential hazard (IP), the greater its radius of influence (R). Figure 4.2 represents the hypothesised

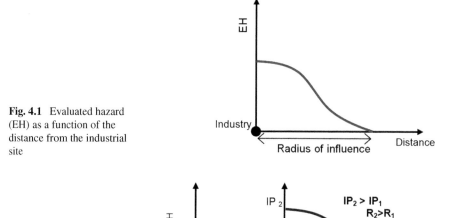

Fig. 4.1 Evaluated hazard (EH) as a function of the distance from the industrial site

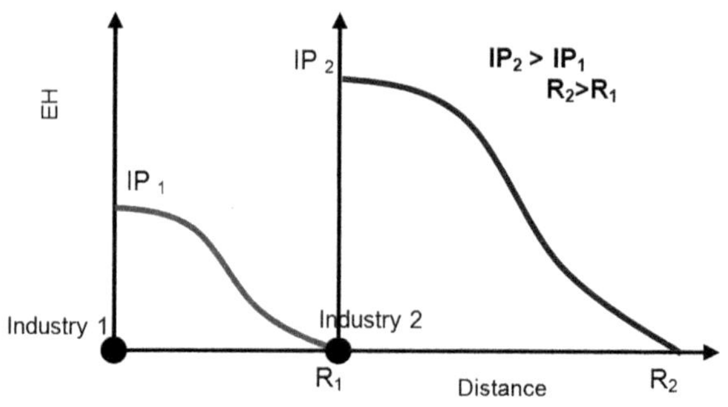

Fig. 4.2 Radius of influence of hazard (R) as a function of maximum EH

4 Evaluating the Firm's Environmental Hazardousness: Methodology

relation. The figure compares the curves of variation of evaluated environmental hazard generated by two Industries. Industry 2 has a larger maximum potential hazard (IP) than Industry 1. Therefore, the Radius of Influence of Industry 2 is also larger than the Radius of influence of Industry 1.

H3: The effect of industries whose areas of influence overlap must be taken into account in order to calculate the total hazard at a given point (EH). The hypothesis is represented in Fig. 4.3.

In a first step the industries are classified in three levels of potential hazard: third category, second category or first category. Third category industries have the higher level of potential hazard and extend their influence over a greater area than second category ones, while first category industries have the lower level of potential hazard and the smallest area of influence.

According to Williamson et al. (1999), the function best suited to represent the phenomenon with the above-mentioned hypotheses is the so-called kernel density. This function has an entrance parameter consisting of a layer of points and a search radius that is applied to each point. The result is a raster that covers the whole area of analysis.

The aim of calculating the kernel density is to estimate how the density of events varies over an area of study, based on a known pattern of points. The advantage of this method is that it translates complex patterns of points onto a smoothed surface that is easier to interpret. Conceptually, this function interpolates a smooth surface fitting for values given at irregularly distributed points. (In simpler terms it creates a smooth curved surface over each point, in this case each industry). The value of this surface is higher where it coincides with the point itself, diminishing as the distance from the point increases and reaching the value 0 when the distance is

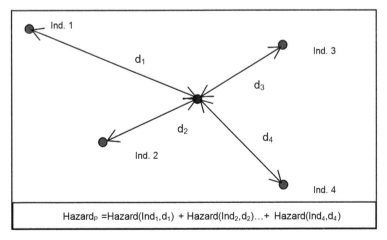

Fig. 4.3 Cumulative hazards at a given point calculated by totalling the impact of each industry as a function of its distance from that point

equal to the radius of influence, after which it maintains the value 0. The volume under the surface (i.e. the integral of the function) is equal to 1. The kernel function is a quadratic approximation to the normalised Gaussian distribution curve, as in the equation:

$$q = k\left[1 - \left(\frac{r}{R}\right)^2\right]^2$$

When $0 \leq r \leq R$; and $q = 0$ when $r > R$

Where q is the value of the kernel function, r the distance from the point to the analysed cell, R the radius of influence and the scale factor k is

$$k = 3/\pi R^2$$

Once the function is applied, a "bell" is obtained for each point as shown in Fig. 4.4.

This function presents suitable characteristics that make it representative of the phenomenon that is analysed, in this case risk. The maximum value coincides with the point that represents the industry, and the values then fall at greater distances from that point, eventually reaching zero at a distance where the industry no longer has any influence, which agrees with hypothesis 1.

Though it would have been easier to calculate a linear distribution, the kernel function gives greater values close to the point, which is desirable as it maintains the highest values of risk in the immediate proximity of the industry. This is illustrated in the comparison of kernel and linear functions in the Fig. 4.5.

The quadratic function is greater than the linear one until a value of approximately 60% of the radius R.

The values in each cell of the raster of the area of analysis are calculated by adding the values of the kernel surfaces that overlap at that point, thus fulfilling the requirements of hypothesis 3.

A weighting factor can also be aggregated for each point, which is equivalent to multiplying the value of the function at each point by said factor.

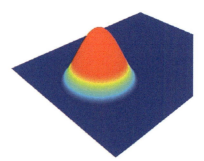

Fig. 4.4 3-D image of the bell of the kernel density function

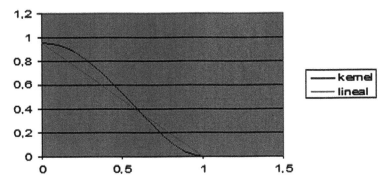

Fig. 4.5 Comparison of the values given by the kernel and linear functions

In our case the Level of Environmental complexity (LEC)² value of each industry will be used as the weighting factor, i.e. the incidence of each industry will be increased as a function of the LEC (hypothesis 2).

Another aspect to be considered is the maximum value that the function reaches, which occurs when it coincides with the point, i.e. when $r = 0$. In this case, the value of the function q is obtained in the same way as the scale factor k, that is:

$$\text{When } r = 0, \ q = k = 3/\pi R^2$$

This means that the maximum value of potential hazard obtained is a function of the radius of influence R adopted. As this is not convenient, a scale factor is applied in order that the final result is a potential hazard value of 1 where it coincides with the industry and of zero at the limit of the radius of influence. To do so the surface is divided by the value of k for each radius. The value thus obtained is then weighted by the LEC, and so the final result will be equal to the LEC when it coincides with the industry and zero at the limit of the radius, as shown in Fig. 4.6.

One of the major difficulties is to determine the value of the radius of influence R to be applied. This value is always subjective, as to affirm that the influence of an industry reaches a certain radius and not another one is necessarily subjective.

Moreover, the value of R has a strong influence on the shape of the final surface. As illustrated by Fig. 4.7 for the same pattern of initial points, high values of R give smoother surfaces, while lower values give more abrupt surfaces with "peaks".

One of the main weaknesses of the method is the arbitrary nature of the selection of the value R. This value, as we have seen, has a great bearing on the degree of

---

[2] In Buenos Aires Province, LEC values are assessed by the Environmental Agency which classifies industries into categories (Type 1, Type 2, Type 3) according to their increasing LEC. However, LEC values were not available for other provinces in Argentina, nor for Spain and Bolivia. Therefore it was necessary to develop an alternative methodology that allows calculation of the weighting factor and categories. The methodology calculates level of complexity of each industry using an algorithm that estimates factors of emission per industrial sector (See Chapter 6).

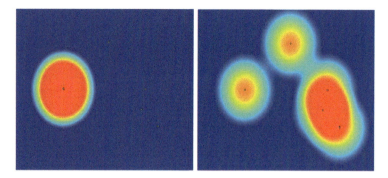

Fig. 4.6 Kernel function applied to a single point and the sum of the effects of several points

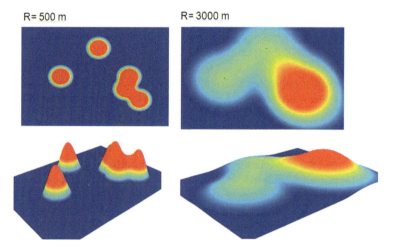

Fig. 4.7 The effect of radius R on the surfaces, given the same initial points

smoothness of the final surface. Although there are no clear parameters to estimate this parameter, several alternatives have been assayed. ESRI Inc., the company that has produced the software, propose a value equal to the lowest dimension of the area of study divided by 30, which does not appear to have any statistical basis. Other authors (Bailey and Gatrell cited by Williamson 1999) propose a formula of the following type:

$$R = 0.68 \, N^{-0.2}$$

This is based on the density of points, but it does not take into account the distance between them, and the nature of the coefficients adopted is not clear enough. One way to overcome these limitations is to use the method of the nearest kth neighbour. This method is based on calculating the mean distances to the nearest

4 Evaluating the Firm's Environmental Hazardousness: Methodology

kth neighbours to each point. Where d_{ij} is the distance between point i and its jth neighbour, then the mean distance to the nearest kth neighbours is expressed as:

$$\frac{1}{kn} \sum_{i=1}^{n} \sum_{j=1}^{k} d_{ij}$$

For instance, if $k = 10$, the estimator is the average of the distances from each point to its ten nearest neighbours. The value of k must be adopted by the analyst in order to specify the desired degree of smoothness on the resulting surface. Small values of k give smaller values of R, and therefore surfaces that are less smooth and more abrupt, while greater values of k give greater values of R and consequently more smooth surfaces.

This way of calculating the radius R is considered better than the previous ones, as it takes into account the distances between the points; in this way the value of R reflects better the spacing and distribution of the points instead of the size of the area studied (Williamson et al. 1999).

To calculate the value of R in the present case, a script in Avenue language was used. This script proposes using a defect value of $k = 30$.

The following step was to analyse diverse alternatives based on the available data. First, the mean distance was calculated with $k = 20$, $k = 30$ and $k = 50$ for the industries as a whole and for each of the categories. Bearing in mind the variations obtained, the suggested value of $k = 30$ was adopted. As a higher category corresponds to greater potential hazard, different values of R were taken depending on the category, and the final distances or radiuses of influence used in the calculation were rounded up as follows in Table 4.1[3]:

The weighting factors to obtain $q = 1$ at $r = 0$ are expressed in Table 4.2.

The functions to be applied, therefore, depending on the parameters adopted, can be seen in Fig. 4.8.

Table 4.1 Values of R adopted depending on the category

Category	Radius of influence R
Second	1,500 m
Third	3,000 m

Table 4.2 Weighting factors

Category	Radius of influence R	Value $k = \frac{3}{\pi R^2}$
Second	1,500 km	0.424413
Third	3,000 km	0.10610

---

[3] First category industries are considered harmless; therefore they are not generators. Second category includes small and medium size generators, Third category includes large generators.

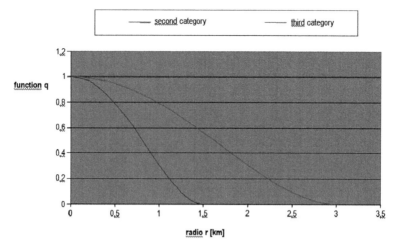

Fig. 4.8 Density function based on the distance from the point and the category, adjusted in order that $q = 1$ at $R = 0$

2.1 Alternative Methodologies Used to Address Paucity of Data

A limitation of the method for calculating *industrial hazardousness* developed in this chapter is that it requires as initial data the geographical location of each particular plant (coordinates of latitude and longitude) and a parameter that allows us to associate each plant with a potential impact. However, many national censuses do not include these types of data. In general, the information that is provided by official sources in each administrative unit is either aggregated (total number of firms by type of industry per given administrative unit); not homogeneous across the whole country; does not include a measurement of impact or covers only a sample of firms. In Ibero-America, only Spain, Chile, Costa Rica and Brazil provide homogeneous information on plant location and type of industry across administrative units. In turn, complete data on emissions per plant or level of impact per plant is available in very few countries (i.e., Chile).[4] Actually, the only plant-related information collected in most countries is plant size (number of employees) and type of industry. In Argentina, for instance, databases containing the geographical location and impact of industries were only available in Buenos Aires province. For this reason, to assess the distribution of risk in the whole country, it was necessary to develop an alternative methodology that (a) allows calculation of accumulated hazardousness of industrial activities by administrative unit when data from individual industries is not available but aggregated data is reported by administrative unit; and (b) allows the use of data on number of employees and type of industry to estimate environmental impact when data on environmental impact is not available.

---

[4] Mexico records emissions of conventional pollutants, sector of production and number of employees for approximately 6000 plants. However, no information on plant location is available.

The methodology calculates potential hazardousness using a model that estimates factors of emission per industrial sector. The model builds on work developed in the World Bank by Dasgupta and Wheeler[5] (2001) to account for specific industry conditions in Latin-America. The algorithm used in our model stresses the importance of incorporating the effect of emissions of small and medium-sized pollutant firms and provides annual polluting particulate emission coefficients of contamination per industrial. These contamination factors measure the intensity of annual emissions of contaminant particles per employee and they vary according to the industrial sector (as different industrial categories are similar to one another in terms of production processes and technologies) and the size of the firm: small (1–20 employees), medium (21–100 employees) and large (over 100 employees). The methodology – applied in Chapters 5 and 6 – is explained in more detail in Chapter 6. It can be used to spatially assess industrial hazardousness in any country where number of employees' and type of industry are consistently measured for all industrial plants – for instance Mexico, Brazil, Peru, and Venezuela – even if such information is only available on an aggregated manner per administrative unit (i.e. total number of employees in the textile industry in the municipality of Olavarria). The more disaggregated data/smaller administrative unit, the more accurate spatial distribution of potential hazardousness.

3 Conclusion

Large urban agglomerations in developing Latin American countries present urban areas in which industries merge with residential zones, in many cases characterised by scenarios of social vulnerability. These vulnerable residential areas are therefore at potential risk depending on the proximity of the dwellings to the industrial installations. The efficient application of policies intended to mitigate possible hazards by authorities or firms requires the creation of categories of potential risk and identifying those areas at greatest potential risk. This allows better allocation of resources which are often in short supply. In most Latin American countries there are legal mechanisms to assess the environmental complexity of each industry, but there is no methodology that extends this assessment spatially to the areas in which industries are established. Similarly, the indicators of environmental impact created by international organisms only provide information on a national scale, and they are therefore unsuitable for the identification of areas at risk.

This chapter tries to fill that gap by proposing methodological tools to diagnose risk areas using geographic information systems. The methodology assesses hazardousness based on the characteristic environmental impacts of the productive

---

[5] The algorithm produced by Dasgupta and Wheeler (2001) is based on mean real values of pollution emitted per province. However, these means come from aggregate data that do not take into account regional and local variations due to factors of regional/local governability. Consequently, on applying this algorithm to individual firms we assume the firm under analysis follows the behaviour of the average firm, thus we are estimating a potential rather than real hazardousness.

processes of each industry and the cumulative effect generated by an agglomeration of industrial activity. This allows the researcher to locate areas of potential hazard due to the accumulation of the effects of small and medium-sized generators/stressors in close proximity to one another, which would not have been identified using methodologies based only on the size or potential impact of individual firms. These areas of potential hazard are superposed on areas of social vulnerability, and where the two overlap we consider that there are areas of potential risk.

The chapter presents a methodology allowing the spatial identification of situations of potential accumulated high hazardousness with an accuracy of 100 metres. This allows the assessment of place-bound "individual threat" at the households block level (the potential threats menacing an individual inhabiting a household in a given block). Such methodology requires as initial data the geographical location of each particular industry (coordinates of latitude and longitude) and a parameter that allows us to associate each industry with a potential. Since our fieldwork revealed that such data is not always readily available in Latin America, the chapter also outlines an alternative methodology to address paucity of data (described in more detail in Chapter 6). Such methodology allows the use of data on number of employees and type of industry to estimate a plant's potential environmental impact. Moreover, it allows the diagnosis of risk areas taking into account the impact of small and medium generators, even when the information about industrial activity is available only in an aggregated manner for a given administrative unit. Although this second methodology does not lead to identification of "individual threat", it provides an introductory assessment of administrative units where situations of high individual risk may exist, thus acting as wake-up call for policy-makers.

Mapping the data to assess situations of hazardousness arising from cumulative negative impacts of all potentially polluting firms, regardless of its size, is an innovative approach. It allowed the identification of hazardousness due not only to large industries, but also to geographical clusters of numerous small industries whose hazardousness was insignificant individually (and therefore less regulated or controlled by policy-makers following Pareto approaches[6]), but whose combined emissions may have constituted a greater threat than that of a single firm.

References

Aceves-Quesada, F., Díaz-Salgado, J., & López-Blanco, J. (2006). Vulnerability assessment in a volcanic risk evaluation in Central Mexico through a multi-criteria-GIS approach. *Natural Hazards,* 40(2), 339–356.
Bankoff, G., Frerks, G., & Hilhorst, D. (2004). *Mapping vulnerability: Disasters, development and people*. London: Earthscan.

---

[6] The Pareto Principle applied to industrial risk assessment broadly states that the top 20% worst polluters of any population of industries cause 80% of the total population's negative effect.

Basta, C., Neuvel, J. M. M., Zlatanova, S., & Ale, B. (2007). Risk-maps informing land-use planning processes. A survey on the Netherlands and the United Kingdom recent developments. *Journal of Hazardous Materials, 145*, 241–249.

Been, V., & Gupta, F. (1997). Coming to the nuisance or going to the barrios? A longitudinal analysis of environmental justice claims. *Ecology Law Quarterly, 24*, 1–56.

Bosque Sendra, J., Díaz Muñoz, M., Gómez Delgado, M., Rodríguez Durán, A., & Rodríguez Espinosa, V. (2000). Sistemas de Información Geográfica y Cartografía de Riesgos Tecnológicos. El caso de las Instalaciones para la gestión de residuos en Madrid. *Industria ymedio ambiente, 1*, 315–325.

Bosque Sendra, J., Díaz Castillo, C., Díaz Muñoz, M., Gómez Delgado, M., Gónzalez Ferreiro, D., Rodríguez Espinosa, V., et al. (2004). Propuesta metodológica para caracterizar las áreas expuestas a riesgos tecnológicos mediante SIG. Aplicación en la Comunidad de Madrid. *GeoFocus (Artículos) 4*, 44.

Bowen, W. (2002). An analytical review of environmental justice research. What do we really know? *Journal of Environmental Management, 29*, 3–15.

Christou, M. D., & Mattarelli, M. (2000). Land-use planning in the vicinity of chemical sites: Risk-informed decision making at a local community level. *Journal of Hazardous Materials, 78*, 191–222.

Cozzani, V., Bandini, R., Basta, C., & Christou, M. D. (2006). Application of land-use planning criteria for the control of major accident hazards: A case-study. *Journal of Hazardous Materials, 136*, 170–180.

Dasgupta, S., & Wheeler, D. (2001). Small plants, industrial pollution and poverty. In R. Hillary (Ed.), *Small and medium-sized firms and the environment* (pp. 289–304). Sheffield: Greenleaf Publishing.

EPA – Environmental Protection Agency. (1998). Risk Assessment Forum. *Guidelines for Ecological Risk Assessment*, EPA/630/R-95/002F.

Fairhurst, S. (2003). Hazard and risk assessment of industrial chemicals in the occupational context in Europe: Some current issues. *Food Chemical Toxicology Journal, 41*, 1453–1462.

Lara-Valencia, F. A., Harlow, S., Lemos, M. C., & Denman, C. (2009). Equity dimensions of hazardous waste generation in rapidly industrialising cities along the United States-Mexico border. *Journal of Environmental Planning and Management, 52*(2), 195–216.

Lirer, L., & Vitelli, L. (1998). Volcanic risk assessment and mapping in the Vesuvian area using GIS. *Natural Hazards, 17*, 1–15.

Malczewski, J. (2006). GIS-based multicriteria decision analysis: A survey of the literature, International Journal of Geographical Information *Science, 20*(7), 703–726.

Meyer, V., Scheuer, S., & Haase, D. (2009). A multicriteria approach for flood risk mapping exemplified at the Mulde River, Germany. *Natural Hazards, 1*, 17–39.

Sengupta, S., & Patil, R. S. (1996). Assessment of population exposure and risk zones due to air pollution using the geographical information system. *Computers, Environment and Urban Systems, 20*(3), 191–199.

Ulberich, A. (2000). Niveles de Riesgo Ambiental Derivados de la Actividad Industrial. Estudio de Caso. Ciudad de Tandil, Buenos Aires, Argentina. *Primer Congreso de la Ciencia Cartográfica y VIII Semana Nacional de la Cartografía*. Buenos Aires.

US GAO. (1983). *Siting of hazardous waste landfills and their correlation with racial and economic status of surrounding communities*. Washington, DC: United States General Accounting Office.

Williamson, D., McLafferty, S., Goldsmith, V., Molenkopf, J., & McGuire, P. (1999). A better method to smooth crime incident data. *Revista ArcUser*, January–March, Chicago: ESRI Press.

Chapter 5
The Case of Bolivia

Luis Augusto Ballivián-Céspedes, Yolanda Bueno-Cachadiña, and Sergio D. López

Abstract This chapter evaluates risk, social vulnerability and industrial hazardousness in Bolivia applying the methodology described in Chapters 3 and 4. As well as presenting aggregated risk results at the departmental level, it provides a more detailed analysis for the municipalities of Santa Cruz and Sucre. The results show high levels of both vulnerability and industrial hazards, especially in the departments with highest economic development. The chapter also draws attention to the need for developing urban planning actions oriented towards a positive evolution of the management of these hazards.

Keywords Environmental risks · Industrial hazardousness · Bolivia · Sucre · Santa Cruz

1 Background: Bolivian Context

The Plurinational State of Bolivia is located at the heart of South America, between longitudes 57° 26′ and 69° 38′ and latitudes 9° 38′ and 22° 53′. Without coastline, it borders to the north and northeast with Brazil, to the northwest with Peru, to the southeast with Paraguay, to the south with Argentina and to the west and southwest with Chile.

With a surface area of 1,098,581 km^2, it is the fifth largest South American country after Brazil, Argentina, Peru and Colombia. It possesses a wealth of geographical diversity, from its lowest point close to the Madera river in the Pando department in the Amazon jungle at 74 m.a.s.l., to Nevado Sajama its highest peak in the Cordillera Occidental at 6,542 m above sea level (m.a.s.l).

The Bolivian territory is divided into three geographic zones:

- *The Andean zone covers* some 307,000 km^2, accounting for 28% of the country's surface area. At an altitude of over 3,000 m.a.s.l. between the Cordillera Occidental and Cordillera Oriental mountain ranges, it includes some of the

---

L.A. Ballivián-Céspedes (✉)
Colegio Pestalozzi, Padilla 174, Sucre, Bolivia
e-mail: auballi@entelnet.bo

highest peaks in America. This area has the lowest temperatures in the country with mean values of 5–10°C.
- *The sub-Andes zone* covers the area between the Altiplano and the eastern plains at altitudes of between 1,000 and 3,000 m.a.s.l. It accounts for 13% of the country's surface area and includes the valleys and yungas (tropical forests). It has a warm, dry climate with mean temperatures of 15–25°C.
- *The plains (llanos)* account for the remaining 59% of the territory and are located to the north of the Cordillera Oriental. They comprise the plain, the low plateau and the jungle. Here the climate is hot and humid, with mean temperatures of 22–25°C.

Bolivia possesses abundant renewable and non-renewable natural resources. It ranks as the sixth country in the world as far as tropical rainforest resources are concerned, third in the American continent behind Brazil and Mexico regarding forests, seventh in the world regarding biodiversity, second in South America regarding gas reserves, and it has major reserves of minerals such as zinc, tin, silver or lithium.

Bolivia is a territory of great diversity and is among the top ten countries with greatest richness of vertebrate species, it is fourth in the world regarding richness of butterflies and sixth regarding species of birds. The country has 14 ecoregions, 199 ecosystems, some 14,000 plant species, 134 timber-yielding species, over 2,600 species of higher wildlife, over 50 autochthonous species and over 3,000 varieties of medicinal plants.

However, these resources are under threat from permanent processes of degradation due to demographic pressure, deforestation, and burning of grasslands, selective extraction of species, illegal hunting and productive activities that contaminate due to the consumables and industrial processes employed, the technological level and the environment in which they are developed.

From an administrative point of view, the constitutional capital of Bolivia is the city Sucre, while La Paz is the seat of government. The political and administrative structure of the country consists of 9 departments, 112 provinces, 314 municipalities and 1,384 cantons. See Table 5.1: General data on the departments of Bolivia.

Table 5.1 General data on the departments of Bolivia

Department	Surface area (km^2)	Capital	Altitude (m.a.s.l.)
Chuquisaca	51,524	Sucre	2,790
La Paz	133,985	La Paz	3,640
Cochabamba	55,631	Cochabamba	2,558
Oruro	53,588	Oruro	3,709
Potosí	118,218	Potosí	4,070
Tarija	37,623	Tarija	1,866
Santa Cruz	370,621	Santa Cruz de la Sierra	416
Beni	213,564	Trinidad	236
Pando	63,827	Cobija	221

5 The Case of Bolivia

The population of Bolivia increased from 2,704,165 inhabitants in 1950 to 8,274,325 in 2001 (National Institute of Statistics, 2001); over this period the urban population increased from 26 to 62% of the total population as a result of massive migration to the cities. By 2010 the population was estimated at 10,426,154 inhabitants (National Institute of Statistics, 2010).

According to data of the National Institute of Statistics, in 2007 Bolivia was the 113th country in the world as far Human Development is concerned, just above such countries as Guyana, Nicaragua, and Haiti in the Latin American context.

According to data of the Fundación Jubileo (2010), in 2008 the rate of moderate poverty was 59.25%, while that of extreme poverty was 32.71% of the population. Most of the poverty was concentrated in rural areas, which account for one third of the Bolivian population and where 53.31% of the population live in conditions of extreme poverty. In such areas, approximately two million people live on less than one dollar a day. The rate of urban unemployment is 8.0%.

The gross domestic product in 2009 was 121,726,745 thousand bolivianos (National Institute of Statistics, 2010), and the main economic activities comprised services of public administration (11.92%); manufacturing industries (11.62%); agricultural, livestock, forestry, hunting and fishing sectors (11.15%); financial entities, insurance and real estate (8.74%) and metallic and non-metallic minerals (7.94%).

According to the Manufacturing Industries Survey (Instituto Nacional de Estadística (National Institute of Statistics), 2001), approximately 1,500 legally established companies made up this sector, and most of these (almost 80%) were located in the cities of La Paz, Cochabamba and Santa Cruz; over 60% of these companies employed 5–14 workers and they focussed in the main on the manufacture of furniture, printing, bakery products, the manufacture of plastic products and the manufacture of clothing.

On the whole, Bolivian industry manufactures products of little aggregate value and employs relatively unskilled workers. Few firms develop economies of scale or invest in improving their productive processes. The vast majority of production is intended for the domestic market, since Bolivia has not yet generated international industrial networks due to its size, domestic market focus and low level of development.

As far as the environment is concerned, little interest was shown in this issue until the 1990s (Escobari, 2003). This means that the country does not possess sufficient information to assess the magnitude of industrial impacts correctly. There is a shortage of studies on the effects of contamination, and the few studies that do exist do not coincide in their aims or interests.

Bolivia's ecological problems do not differ greatly from those of other Latin American countries. Among the main issues we should include loss of vegetation, soil erosion, deforestation, burning of pasture and woodland, loss of biodiversity, indiscriminate use of agrichemicals, excessive grazing, contamination of waters due to mining, and lack of industrial treatment of urban waste.

Industrial management in Bolivia was developed for the most part over the 1990s. At that time a general legal framework was adopted and specific rules were drawn up for the concession of industrial licences for polluting activities. Industrial norms were passed to regulate the industrial quality of the hydrocarbons and mining sectors. At government level a ministry was created to deal with industrial issues. This legal framework was established by the Law for the Environment (Ley del Medio Ambiente) and its corresponding regulations; the Law ratifying the framework convention on climate change (Ley que Ratifica la Convención Marco sobre el Cambio Climático); industrial regulations for the hydrocarbons, mining and manufacturing sectors (los Reglamentos Ambientales para los Sectores Hidrocarburos, Minero y Manufacturero); the norm to regulate renewable natural resources (el Reglamento para el Sistema de Regulación de Recursos Naturales Renovables) and the regulation on industrial management of ozone-depleting substances (Reglamento de Gestión Ambiental de Sustancias Agotadoras de Ozono). Nonetheless, despite the progress made in formulating these regulations, it remains for the authorities to guarantee their enforcement and to adapt them to the changes that the country has undergone.

In Bolivia, the industrial sector is the main responsible for generating solid, hazardous waste, rivers' contamination and air pollution (Escobari, 2003):

- *Agriculture.* Among the industrial impacts of this sector are those derived from using or generating contaminant products; the use of land and water resources; the use of chemical agents to improve productivity that have considerable effects on health and which generate toxic and greenhouse-effect gases due to "chaqueos" (burning of land for agricultural and livestock purposes).
- *Mining.* The industrial problems caused by the medium-scale mining sector are mainly due to not taking precautions to avoid soil and hydric contamination; this sector consumes huge quantities of water, most of which returns to its natural source untreated, releasing large amounts of mercury, for example, due to lead and gold mining; industrial externalities are due to the generation of different pollutants during extraction processes; tail dams, waste disposal and the processes involved in closing the mines. Traditional small-scale mining is an extremely dirty and barely feasible process owing to the lack of sources of finance, non-competitive production costs, insufficient investment and reinvestment, low grade of minerals and obsolete technology. Although the impact of each firm may be slight, the great number of cooperativists means that there is a considerable industrial impact which is proportionally greater than that of medium-scale mining.
- *Energy industry.* The most contaminating activities in this sector are linked to the exploitation, transport and refining of oil and natural gas.
- *Manufacturing industry.* The industrial impact of manufacturing residues varies greatly depending on the industrial activity, the raw materials and the processes used. Although the proliferation of non-legal firms makes it difficult to obtain reliable data, this sector has a great impact due to the manufacturing processes used, limited access to technology and the low levels of income. The worst contaminators are metallurgical firms, those in the industrial mineral subsector, such as cement manufacturers, tanneries and the food industry.

5 The Case of Bolivia

The analysis of this chapter focus on two cities with radically different levels of industrial development: Sucre and Santa Cruz.

Sucre, founded in 1538, is the constitutional and historical capital of Bolivia and the capital of the department of Chuquisaca. It is a colonial city, seat of the renowned university San Francisco Xavier, founded in 1624, the fourth oldest in America. Sucre was the only capital of Bolivia between 1825, when the country obtained independence from Spain, and 1889. In 1889, a pact between the conservative party and the liberal avoided civil war by moving the presidency and the Congress to La Paz, which became the de facto capital of the country. However, Sucre maintained its status of "constitutional" capital of Bolivia and seat of the judicial power. Nowadays, Sucre is a university and administrative city with a population of 306,754 inhabitants and limited industrial development.

Santa Cruz de la Sierra is the capital of the Autonomous Department of Santa Cruz. It is a dynamic emergent city, the largest and most populated in Bolivia, and it is considered the economic and industrial capital of the country. From 10,000 people in 1810 and 18,000 in 1910; its population grew to 57,000 in 1955; 325,000 in 1976; 697,000 in 1992; 1,029,471 in 2001 and an estimated 1,651,436 in 2010. Its population, economy and surface grows so fast that, in the turn of a generation, it has changed from a small village into a vast city which has surpassed the limits of the municipality of Santa Cruz de la Sierra, and which newer suburbs have connected her with the neighbouring municipalities of La Guardia, Cotoca, Warnes, Montero, El Torno and Porongo. Its metropolitan area has an estimated population of 2,102,998. Its demographic growth is among the fastest in South America and, at present, it is number 14 in the list of fastest growing cities in the world. In 2006, Santa Cruz de la Sierra became the largest city in Bolivia. Santa Cruz is the city that undergoes the greatest transformations in Bolivia, due to its high growth and migration levels, which demand a permanent search for improvements in infrastructure, health services and education. The city's economic structure is tertiary and informal. The tertiary sector represents 94% of the economic businesses and 85% of occupied people. Informal work market involves 60% of the population.

Santa Cruz de la Sierra lies on the right bank of the river Piraí, which runs north to flow into the river Grande or Guapay, part of the Amazonia basin. It is 416 m above sea level, its topography is flat and its coordinates are 17°48′02″S and 63°10′41″W. The area taken up by the city is 567 km^2, and it has a perimeter of 110.2 km. Santa Cruz de la Sierra alone is greater in extension than La Paz and El Alto put together. The total extension of the metropolitan area of Santa Cruz de la Sierra is 1,590 km^2, making it bigger that Montevideo, Asunción or Brasilia. As the economic driving force of Bolivia and capital of the widest department of the country, Santa Cruz de la Sierra has important road and public services infrastructure and an active business and commercial life. The area has the highest density of industrial facilities in Bolivia. The main sectors within its economy are crude oil, forestry activities, agribusiness and building. Santa Cruz concentrates more than 80% of the agricultural national produce and it contributes to the country's GDP with more than 35%, according to recent year's data. It also owns the country's main airport, which makes it an ideal location for trade fairs, international events and investments. In Santa Cruz there are considerable investments in building, trade, health, fashion, national and international shows, agro industry, and in hotel and catering business.

2 Data Collection and Methodology of Analysis

The case study of Bolivia has been limited mainly by the shortage of available information for both industrial hazardousness and social vulnerability. The lack of information at greater levels of disaggregation has limited the mapping at the following levels:

- *Industrial hazardousness*. A map has been elaborated with aggregate values for Bolivia's capital cities of departments and the maps of hazardousness for two cities that presented great differences: Santa Cruz de la Sierra, which has a very high level of industrial hazardousness, and Sucre, whose level is very low.
- *Social vulnerability*. An aggregate map has been made for the capital cities of departments. In the cases of Sucre and Santa Cruz de la Sierra, given the lack of disaggregate information, approximate maps were elaborated based on a qualitative analysis.
- *Evaluated risk*. Based on the maps of the previous two variables for the capital cities of departments an aggregate map of those cities has been drawn. In the cases of Sucre and Santa Cruz de la Sierra, overlapping the maps of industrial hazardousness and social vulnerability gave rise to maps of evaluated (potential) risk for those cities.

The indices were calculated as follows:

- *Social vulnerability*. Nine indicators were used to calculate social vulnerability: single-parent households, population of 14 years of age or less, population of 65 years of age or over, illiteracy rate, population without running water, population without sewage treatment, unemployment rate, population without access to social security and without their basic needs covered. These data were taken from the 2001 Population and Housing Census of the National Institute of Statistics (Censo de Población y Vivienda del Instituto Nacional de Estadística, 2001).

These indicators were scaled using the method of natural breaks, which is a method for classifying data and determining the best arrangement of the values in the different classes in a scale of 1–5, where 1 = very low, 2 = low, 3 = medium, 4 = high and 5 = very high. The index of social vulnerability was obtained by rescaling the total of the nine values on the same scale.

The qualitative maps of the cities of Sucre and Santa Cruz de la Sierra considered a value for the different areas of the city on a scale of 1–3, where 1 = low, 2 = medium and 3 = high.

- *Industrial hazardousness*. For the aggregate map of hazardousness of cities the index was determined on the basis of the number of companies present in each city in each of the different industrial activities, the average number of employees and a coefficient of the annual emission of particles which varied depending on the sector and size of the company. These data were used to calculate the likely annual emission of particles of each sector, the sum of which would represent the quantity of particles emitted annually in each city.

These values were scaled using the method of natural breaks in order to determine the indices of industrial hazardousness on a scale of 1–5, in a similar way as described above for social vulnerability.

To prepare the maps of hazardousness for the two cities chosen as case studies, the information compiled included: the name of the firm, the city and department of its location, its address, the sector to which it belongs and the average number of employees in that sector.

This information was obtained from the manufacturing industries survey (Instituto Nacional de Estadística, 2001), the United Nations Development Programme (UNDP), the analysis unit of social and economic policies (Unidad de Análisis de Políticas Sociales y Económicas (UDAPE), 2003), the databases of the Bolivian Chambers of Industry and Commerce (Cámaras de Industria y Comercio de Bolivia), the Bolivian Federation of Private Entrepreneurs, the Chamber of Exports, Embassies and bilateral Chambers of Commerce, business guides published in the course of the last year, directories of firms published in the yellow pages of each of the capital cities, indexes of firms registered in the Town Halls and information provided by the Administration of Industrial Parks in Santa Cruz and Cochabamba.

- *Combined risk of industrial hazardousness and social vulnerability*. This index was obtained from the total sum of the indices of social vulnerability and industrial hazardousness for each city on a scale of 1–5 using the method of natural breaks.

The aggregate maps of capital cities of departments were drawn up based on the values determined by the indices.

For the analysis of industrial hazardousness of the two cities selected, Sucre and Santa Cruz de la Sierra, each of the firms was located on a map using specific software for GIS, and the maps were drawn up on the basis of the information obtained and the methodology explained in Chapter 4.

The compilation of data for the Bolivian industrial sector provided the information shown in Table 5.2 for the different department capitals.

The map of social vulnerability for the cities of Sucre and Santa Cruz de la Sierra were drawn up using a qualitative evaluation by a panel of local people,[1] who were

---

[1] The Sucre panels consisted of 6 people each. The first one was comprised by 2 secondary school teachers of social science, 1 student from the final year of high school (baccalaureate), 1 social

Table 5.2 Number of firms per department

City	Number of firms
Sucre	76
La Paz	417
Cochabamba	375
Oruro	73
Potosí	27
Tarija	61
Santa Cruz de la Sierra	550
Trinidad	19
Cobija	11

asked to map the areas they considered to be of high, medium or low vulnerability. The results were then analysed and contrasted by the panel in order to reach a consensus and draw up the final map.

Later, these maps were validated by different panels to which they were presented, such panels were also asked to revise and, if necessary, amend the maps.

Among the most important limitations that were encountered in the course of the research we should highlight the following:

- The information dates back to 2001, the date of the last census in Bolivia.
- For the estimation of industrial hazardousness no institution or company, either public or private, has complete information on the country's industrial entities that is freely accessible to the public. The information provided by the National Institute of Statistics (Instituto Nacional de Estadística) is of a very general nature and with a high level of aggregation.
- In Bolivia there are a great number of firms which are not registered with any institution. Therefore the industrial guides published by the Chamber of Industry and Commerce or the Federation of Private Entrepreneurs (Federación de Empresarios Privados) are incomplete.
- It has proved impossible to find information on the exact number of employees in each firm. It has only been possible to obtain the average number of employees in firms of the industrial category to which each of the activities belongs in each capital city. No information is available on rural areas, and so it was decided to limit the study to the capital cities.
- The maps of social vulnerability for Sucre and Santa Cruz de la Sierra were drawn up based on the perception of panels. It is therefore a subjective evaluation and can only represent an approximation, as is the case for the map of combined risk.

---

worker, 1 architect, and 1 economist. The panel that validated the map was made up by 2 university students, 2 architects, 1 social worker and 1 university teacher. The panels in Santa Cruz de la Sierra were also constituted by six people each. The first one was made up as follows: 2 doctors, 1 architect, 1 industrial engineer and 2 social workers; the second one included 1 doctor, 1 civil engineer, 1 social worker, 2 university students and 1 business administrator.

3 Industrial Hazardousness in Bolivia

The descriptive map of industrial hazardousness in the capital cities of departments is shown in Fig. 5.1.

The capitals of department that present the greatest industrial hazardousness are Santa Cruz de la Sierra, La Paz and Cochabamba. This coincides with the fact that most of the country's manufacturing industries are established in these three cities (83% of the industrial businesses identified in this research) and that they are responsible for 84% of the total emissions at national level. It is also noticeable that these three cities concentrate almost 30% of the country's population.

Trinidad and Tarija are in an intermediate position, while Oruro, Sucre, Potosí and Cobija are included in the profile of low or very low industrial hazardousness.

Fig. 5.1 Map of industrial hazardousness for the departmental capitals of Bolivia

The enormous difference in environmental hazardousness, between the biggest cities in Bolivia and those intermediate or small, can be better appreciated in Fig. 5.2, where the values obtained from the calculation of particle emission in the industrial sector by employee per capital of department, are represented.

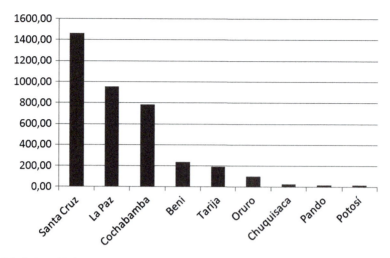

Fig. 5.2 Index of industrial hazardousness (IP) for the departmental capitals of Bolivia

4 Social Vulnerability in Bolivia

For each capital of department, 9 sub-indexes or partial indicators of Social Vulnerability 1–5 were developed in a scale of 1–5 using the natural breaks methodology. The sub-indexes were subsequently added and re-scaled to obtain

Fig. 5.3 Map of social vulnerability for the departmental capitals of Bolivia

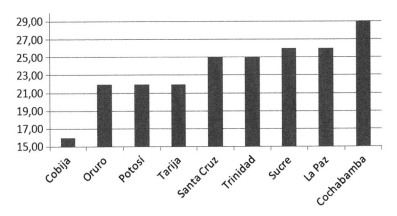

Fig. 5.4 Index of social vulnerability (IVS) for the departmental capitals of Bolivia

the combined Index of Social Vulnerability that identifies each of the capital of department cities mapped in Fig. 5.3.

Next, Fig. 5.4 reveals the following: Cochabamba, Sucre and La Paz present high or very high values; Trinidad and Santa Cruz de la Sierra present intermediate values, whereas Oruro, Potosí, Tarija and Cobija have low or very low values.

5 Combined Risk of Industrial Hazardousness and Social Vulnerability in Bolivia

The superposition of the layers used to create the map shown in Fig. 5.5, reveals the distribution of risks resulting as the combination of the two variables considered, industrial hazardousness and social vulnerability, for the capital cities of each department.

Figure 5.6 shows below that the high and very high risk is centred on the cities of Santa Cruz de la Sierra, La Paz and Cochabamba, where there is on the one hand a higher concentration of Bolivia's industrial firms, and on the other the highest concentrations of population (3,110,029 inhabitants, i.e. 76% of the population settled in capitals of department).

Trinidad is also included in this high risk category due to the combination of high social vulnerability and medium level industrial hazardousness.

Sucre and Tarija are in the medium range of combined risk for a combination of reasons: in most areas both industrial threat and social vulnerability show middle range values, in the remaining areas extreme levels of the previous two indices balance each other out (i.e. high level of vulnerability is balanced by low industrial threat and the other way around). These are cities of intermediate size with relatively little industry.

Potosí and Oruro, whose indices of social vulnerability and industrial hazardousness are low or very low, fall into the category of low combined risk.

Fig. 5.5 Map of combined risk for the departmental capitals of Bolivia

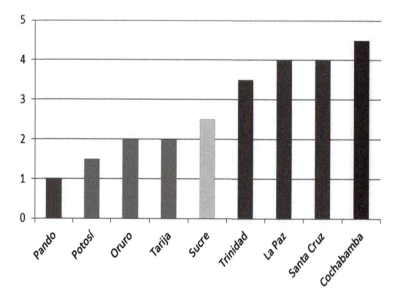

Fig. 5.6 Index of combined risk for the departmental capitals of Bolivia

Finally, Cobija presents very low combined risk, due to the combination of the very low values of both indices.

6 Analysis of Industrial Hazardousness for the Cities of Sucre and Santa Cruz de la Sierra

One of the aims of this project was to analyse the levels of risk in greater detail than municipal level. This proved impossible for the index of social vulnerability due to the aggregate nature of the available information, but it was possible for industrial hazardousness.

Table 5.3 Environmental Hazard Contribution by industrial sector in Sucre

Industrial sector	Particle emission Percentage	Cumulative percentage
Timber	19.26	19.26
Manufacturing of cocoa, chocolate and confectionery	18.89	38.15
Manufacturing of metallic produce for structural use	14.56	52.71
Production, processing and conservation of meat and meat products	6.47	59.18
Manufacturing of cement, lime and plaster	6.38	65.57
Manufacturing of concrete, cement and plaster goods	5.01	70.57
Others	29.43	100.00

Table 5.4 Environmental Hazardousness Contribution by industrial sector in Santa Cruz de la Sierra

Industrial sector	Particle emission Cumulative Percentage	Cumulative Percentage
Timber	22.38	22.38
Production, processing and conservation of meat and meat products	17.51	39.89
Sugar manufacturing	8.49	48.38
Manufacturing of oil refining products	8.05	56.43
Manufacturing of plastic products	7.79	64.22
Manufacturing of timber sheets for veneer; plywood boards; laminated boards; particle boards and other boards and panels	6.06	70.28
Others	29.72	100.00

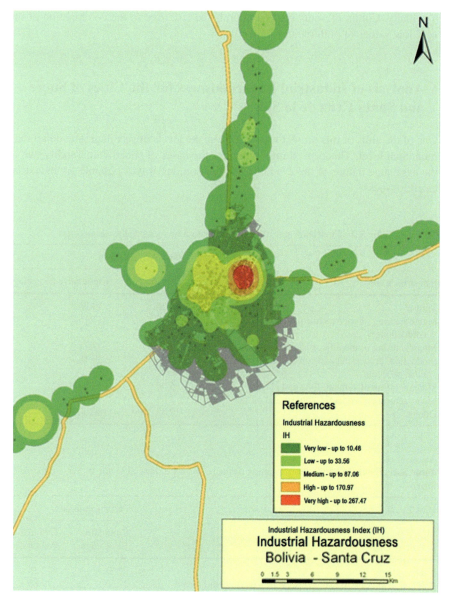

Fig. 5.7 Map of industrial hazardousness for Santa Cruz de La Sierra

5 The Case of Bolivia

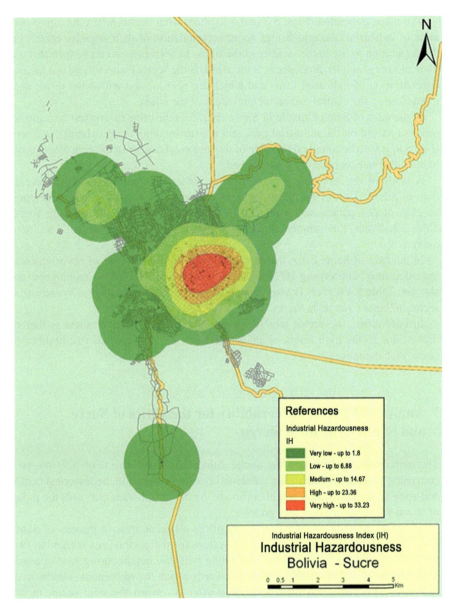

Fig. 5.8 Map of industrial hazardousness for Sucre

The maps generated (Figs. 5.7 and 5.8) show what occurs when the agglomeration of industrial concerns brings about accumulation of their negative effects on the population living in the vicinity of the firms. In both cases it can be seen that the high and very high hazardousness is focussed on the sectors where there is a greater concentration of industrial firms and which are also located within the urban area quite close to the central and residential areas of the cities.

In the case of Santa Cruz de la Sierra (Fig. 5.7) the zone of greatest hazardousness is focused on the industrial park and the surrounding areas, whereas in Sucre (Fig. 5.8) it is to be found on one side of the historical centre, in the area where most industrial concerns have been established.

A study of the industrial sectors by their potential of particle emission reveals that, in Sucre, 70% of the hazardousness is due to six sectors which correspond to 25% of companies and, which, in total, engage 46% of the working force. Table 5.3, depicts Environmental Hazardousness Contribution by industrial sector in Sucre.

In the case of Santa Cruz, the sectors responsible for the 70% of environmental hazardousness, concerning 20% of the businesses and 32% of the employees, are also six. Table 5.4 depicts Environmental Hazardousness Contribution by industrial sector in Santa Cruz de la Sierra.

In both cities, the sector which imposes the greatest hazardousness is that of timber, due to the high levels of annual particle emission and the vast number of employees it engages.

7 Analysis of Social Vulnerability for the Cities of Sucre and Santa Cruz de la Sierra

The map drawn up to reflect the social vulnerability in Sucre (Fig. 5.9) shows a concentric pattern in which least vulnerability is to be found in the historical centre and some of the residential neighbourhoods. As the city spreads outwards the index of social vulnerability becomes greater.

The map showing social vulnerability of the different areas in the city of Santa Cruz de la Sierra (Fig. 5.10) reveals low values for the central area within the two first sectors and part of the third one, along with the neighbouring road towards Viru-Viru airport. These are the sectors towards which the residential suburbs have extended. The medium vulnerability area extends toward the third sector in the Eastern district of the city, and also to the road to Viru-Viru airport. The rest of the city, comprising the peripheral areas beyond the fourth sector, shows the highest levels of social vulnerability.

5 The Case of Bolivia

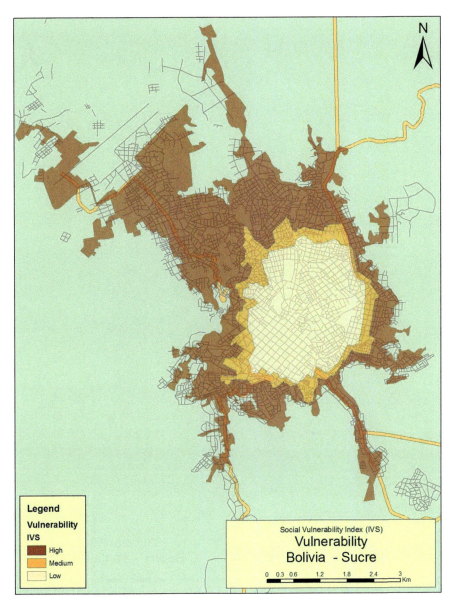

Fig. 5.9 Map of social vulnerability for Sucre

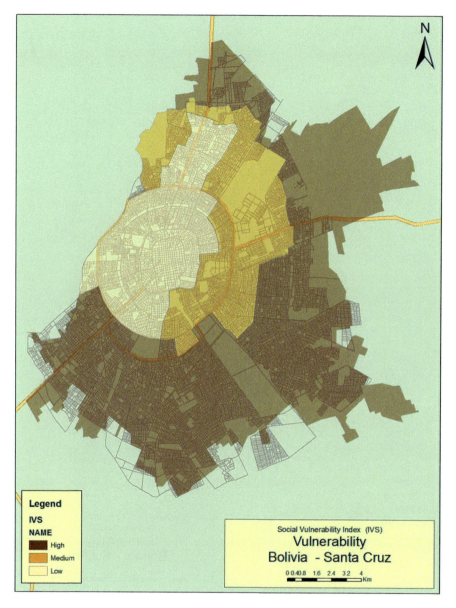

Fig. 5.10 Map of social vulnerability for Santa Cruz de la Sierra

8 Analysis of the Combined Risk of Industrial Hazardousness and Social Vulnerability for the Cities of Sucre and Santa Cruz de la Sierra

Superposing the layers of industrial hazardousness and social vulnerability provides us with a map of distribution of areas in situations of evaluated risk (Figs. 5.11 and 5.12). A *very high risk* zone occurs where an area of high social vulnerability

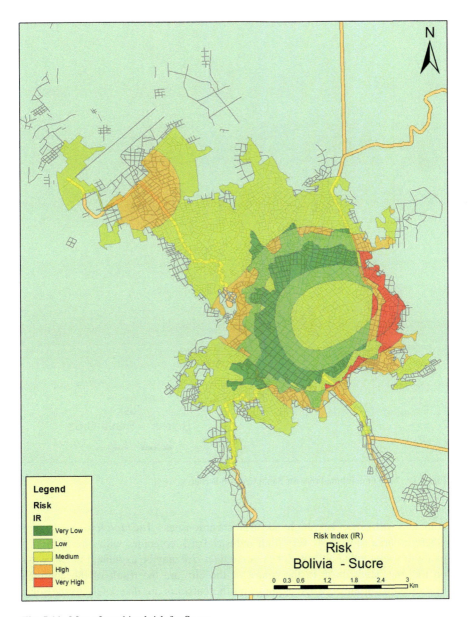

Fig. 5.11 Map of combined risk for Sucre

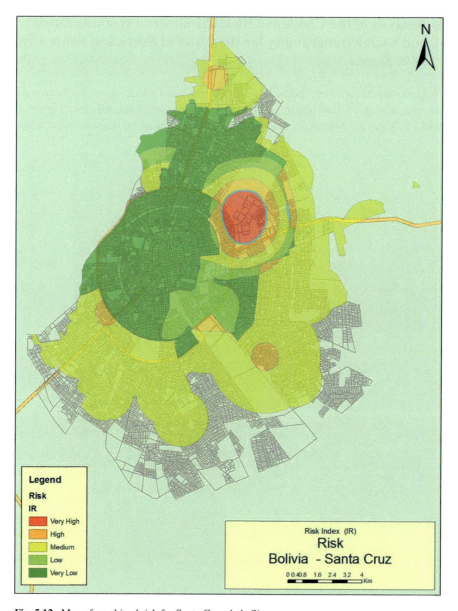

Fig. 5.12 Map of combined risk for Santa Cruz de la Sierra

coincides with one of very high industrial hazardousness. The *high risk* zone appears where an area of high or very high industrial risk coincides with one of high or medium social vulnerability. Zones of *medium risk* may be explained in two ways: on the one hand, those in the centre of the city are the result of the combination of both variables, whereas those on the outskirts are due solely to the social vulnerability factor. The rest of the city presents *low risk* indices (low vulnerability and very low or low threat), with the exception of a small enclave that has a *very low risk* value (low vulnerability and very low threat).

In Santa Cruz de la Sierra it can be appreciated that the highest risk area is the one surrounding the industrial park. High risk sectors are those surrounding the industrial park and the sugar refinery.

The low risk areas are those residential suburbs away from the industrial park and, interestingly, the suburbs where, despite the higher social vulnerability, the combined risk is low due to the inexistence of industrial businesses.

In contrast to Sucre, no sector of the city has been classified as "very low risk".

Using the GIS to elaborate a rough estimation of people living in hot-spots, suggests than 100.000 inhabitants of Santa Cruz and 16.000 inhabitants of Sucre live in areas with high or very high environmental risk. Although the numbers are estimates, these suggest a bias towards the vulnerable in the allocation of environmentally threats.

9 Conclusions and Implications

From the viewpoint of industrial hazardousness, the analysis of the two cities reveals that there is a segment of population that lives or works in the areas of greatest impact. These inhabitants are therefore exposed to the potential adverse effects of the industries, since the cities' rapid growth and expansion have meant that industrial sites are located within the urban radius, coinciding with residential areas.

On the aggregate level of capital cities of departments, the relationship between social vulnerability and hazardousness can be appreciated in that it is precisely those cities that are most "developed" or "progressive" (Santa Cruz de la Sierra, Cochabamba and La Paz) that present the greatest evaluated risk due to the combination of both variables.

Within the range of high and very high risk there is a certain correlation among the three indices. This is tantamount to saying that the greater the population (exposure), the greater the industrial hazardousness (due to greater presence of industries) and the greater the social vulnerability. The same can be said for low or very low risk, except that the opposite occurs.

Nevertheless, there is no clear correlation between the three indices in the cities with medium level of risk, as there may be only medium or high values that are counteracted by low or very low ones.

In the case of Bolivia, data is only available by departments so there is no information disaggregated at the level of census unit, which has made it impossible to obtain exact, reliable results. Nonetheless, the solutions adopted bear testimony to the flexibility of the proposed methodology and its suitability for different scenarios and circumstances.

The results are sufficient to draw attention to a topic which has not been perceived as important in Bolivia and which has not been the object of research, particularly in the depth to which the present project has gone. The analysis carried out in the present work has shown that as there are no tools to assess potential risk, no measures of urban organisation are taken to mitigate the potential adverse effects of industry on the population. From this point of view, the methodology adopted proves its validity by providing not only data, but also maps that allow political and social decision makers to apply policies and take action with a view to mitigating the evaluated risk to which the population is exposed.

References

Escobari, J. (2003). *UDAPE. Unidad de Análisis de Políticas Sociales y Económicas*. Retrieved February 15, 2010, from http://www.bvsde.paho.org/bvsacd/cd29/escobari.pdf
Fundación Jubileo. (2010). *Balance Económico Social.*
Instituto Nacional de Estadística. (2001). *Encuesta de la Industria Manufacturera*. La Paz: INE.
National Institute of Statistics. (2001). Population and Housing Census of the National Institute of Statistics (Censo de Población y Vivienda del Instituto Nacional de Estadística)
National Institute of Statistics. (2010). *Economic Outlook.*
UDAPE – Unidad de Análisis de Políticas Sociales y Económicas. (2003). *Documento de estructura del sector industrial manufacturero 1990–2001.* Retrieved November 17, 2009, from http://www.udape.gob.bo/portales_html/Documentos%20de%20trabajo/DocTrabajo/2003/INDUSTRIA.pdf

Chapter 6
The Case of Argentina

Claudia E. Natenzon, Diego A. Vazquez-Brust, and Sergio D. López

Abstract This chapter presents the results of mapping industrial risk in Argentina and explains the empirical procedures followed to obtain the maps. The chapter is divided into three parts: The first part provides background information about industrial hazardousness in Argentina, while the second one studies the distribution of risk in the country, using the department or municipality as the unit of analysis. The third part presents a case study of the region with the highest concentration of departments/municipalities at high risk: the MABA (Metropolitan Area of Buenos Aires) using the census block group as the unit of analysis. The chapter also explores qualitatively situations of "environmental injustice" and notes that the conclusions regarding the correlation between vulnerability and environmental hazard in the case study differ from those obtained at national level. When the unit of analysis is census block group the spatial distribution suggests an inverse relationship between vulnerability and environmental hazard, where the risk gradient decreases with distance from the city of Buenos Aires as the social gradient of vulnerability increases. Although more detailed studies are required, this result suggests the need to develop indicators including different geographical units of analysis to examine local changes in the distribution of hazard trends.

Keywords Environmental risks · Industrial hazardousness · Argentina · Metropolitan area Buenos Aires · Environmental justice

1 Background: Argentinean Context

Argentina is located at the south of South America, stretching 4,000 km from subtropical landscapes in the North to tundra and sub-polar climate in the far South. Its diverse geography includes both the highest (Cerro Aconcagua) and lowest

---

C.E. Natenzon (✉)
Faculty of Philosophy and Letters, Institute of Geography "Romualdo Ardissone",
University of Buenos Aires, Buenos Aires 1406, Argentina
e-mail: natenzon@filo.uba.ar

D.A. Vazquez-Brust (✉)
The Centre for Business Relationships, Accountability, Sustainability and Society (BRASS),
Cardiff University, Cardiff Wales CF10 3AT, UK
e-mail: VazquezD@cardiff.ac.uk

(Laguna de Carbon) points in the Western Hemisphere. It is the 8th largest country in the world, the second largest in Latin-America (after Brazil) and the largest Spanish speaking country. At present, Argentina is divided into 24 jurisdictions: 23 provinces and the CABA – Autonomous City of Buenos Aires,[1] home to the national government. These provinces are sub-divided into 512 administrative units termed "departments", with the exception of the Province of Buenos Aires, where they are termed "partidos" (see: www.indec.gov.ar).

Despite its size, Argentina has a population of only 40.6 million inhabitants and the lowest population growth rate of Latin America at 1% (ECLAC, 2010). In addition, Argentina is a highly urbanised country and most of its population and economic activity is clustered in the temperate central region, while the vast expanses in the north and south are scarcely inhabited. Only 10% of the population is located in rural areas whereas half of its inhabitants are concentrated in the country's ten largest metropolitan areas. 13 millions live in the Case Study area analysed in this chapter: the greater Buenos Aires metropolitan area (MABA), one of the most populated urban conglomerations in the world. Argentina is the fourth largest economy in Ibero-America (after Spain, Brazil and Mexico) and the second in terms of purchasing power after Spain (24th in the world) (ECLAC, 2010). Argentina is a traditional middle-class country with a well educated work force, high levels of literacy and Human Development Index, low child mortality rate, abundance of natural resources and a diversified economy. The country's three first exports are soy, cars and gold. 10% of the country is fertile land. Agricultural products account for more than 50% of the country's income from exports. However, agriculture represents only 9% of total GDP while manufacturing accounts for 21% and Services for 59% of total GDP (CIA, 2011). A century ago Argentina was one of the worlds' wealthiest countries but it has been doomed by cyclical economic crises since the 1930s depression. The worst of all such crises was in 2001, when a combination of recession, booming debt, currency devaluation and political instability culminated in the country defaulting on its external debt. The crisis left half of the country in poverty and led to situations of social turmoil and institutional collapse. The last decade has seen Argentina in a cycle of sustained growth and sharp reduction of poverty. However, growth has been accompanied by inequalities in the distribution of income, exposure to environmental risks and access to education and infrastructure. Thus, enclaves of extreme poverty and social vulnerability still exist even in the wealthiest areas of the country.

2 Management of Risk and Industrial Activity in the Republic of Argentina

The pattern of environmental degradation in Argentina – i.e. water and air contamination, industrial waste, deforestation, soil degradation – is typical of developing countries with highly concentrated urban populations (Dasgupta and Wheeler 2001;

---

[1] For the CABA the 2001 Census presents disaggregate information for the 21 school districts, but it will not be used on this occasion.

Hochstetler 2002). For all practical purposes formal regulation of industrial activities did not exist until the 1980s despite high levels of industrial pollution. Towards the end of that decade, growing awareness slowly mobilised community and judicial stakeholders into actions and enactment of some basic environmental legislation took place (Hochstetler, 2003). It was nonetheless poorly enforced, with the typical firm relegating matters of the environment to its legal and/or marketing departments.

Two major domestic environmental accidents, an oil spill and volcanic ash damage, spurred the formation of the National Environmental Secretariat in 1991. Its original mandate was to design and enforce "command and control"-type policies that coerced firms into investing in end-of-pipe technologies for the treatment and disposal of residues. At the end of the same decade, the discourse of sustainability began to predominate, stressing the voluntary participation of the industrial community and prioritising actions directed at global climate change. Argentina became a world leader in setting voluntary greenhouse gas targets (CIA 2011) whereas problems of a more local nature related to the contamination of rivers and underground aquifers, the elimination of dangerous residues and the protection of ecosystems were relegated to a secondary role (Hochstetler 2002).

The financial crisis of 2001–2002 held up the institutional advances made in terms of environmental management. Economic and political problems gave rise to generalised disinterest in environmental issues, and this was accompanied by a lack of resources for the application and enforcement of regulations (Hochstetler 2003) and a general lack of direction and leadership (Chudnovsky et al. 2005). Government policies and firms' social responsibility policies became more concerned with programmes to alleviate poverty.

The environment continued to be off the list of social priorities until 2004, when the country started to enjoy a new period of relative development. Although economic growth and alleviating poverty maintained a dominant position on the political agenda, there was resurgence in public concern for environmental risks which was reflected in renewed, though weak, pressure on firms and the authorities to optimise the management of environmental risks. The authorities responsible for these issues received additional powers and greater resources with which to carry out their duty of controlling industrial contaminators (Vazquez-Brust et al. 2010). In turn, Argentinean diplomacy enhanced the country's high-profile in environmental issues, subscribing and actively promoting a plethora of international pro-environmental agreements.[2]

Widespread activism on the part of environmental NGOs had a considerable impact on multinational firms that operate in contaminating sectors. One example worthy of mention is the mobilisations of grassroots activists opposing open-pit gold mining, due to its high environmental impact. Social resistance to "contaminating

---

[2] Environmental Protocol, Antarctic-Marine Living Resources, Antarctic Seals, Antarctic Treaty, Biodiversity, Climate Change, Climate Change-Kyoto Protocol, Desertification, Endangered Species, Environmental Modification, Hazardous Wastes, Law of the Sea, Marine Dumping, Ozone Layer Protection, Ship Pollution, Wetlands, Whaling.

mining" was successful in stopping a multinational investment in the municipality of Esquel (Chubut province), and this action gave rise to regulations restricting open-pit gold mining in six provinces. Vazquez-Brust et al. (2009) and Vives (2006) nonetheless doubt that NGOs will make important progress in Latin America until they are able to communicate effectively in the 'language of business'.

There is evidence that the business community responded in terms of the voluntary management of environmental risks generated by industrial activities. For instance, Argentinean membership of the Global Compact (a UN programme of voluntary environmental management) is increasing, and the numbers of firms that have implemented ISO 14001[3] standards rose from 249 firms in 2002 to 1163 in 2008 (ECLAC, 2010). ISO 14001 certification is now promoted by the Argentinean Chamber of Industry which has awarded the status to over 100 firms per year. Nevertheless, there is still considerable room for improvement. Even in the context of Latin America, Argentina is clearly lagging behind Brazil and Chile (Newell and Muro 2006) and recent statistics show a sharp decrease in the number of ISO 14001 certified enterprises – from 1163 in 2008 to 676 in 2009 (ECLAC, 2010).

3 Data Compilation and Methodology of Analysis

This chapter first analyses industrial risk in Argentina as a whole and then offers a more detailed analysis on the particular case of the Metropolitan Area of Buenos Aires (MABA[4]). The maps presented refer to industrial hazardousness, social vulnerability and evaluated risk. For Argentina as a whole the three maps take the departments as the geographical units of analysis. For the case of the MABA, the maps of evaluated risk and vulnerability use information at the level of the census block group (the smallest unit of dissemination of census data) while the industrial hazardousness map uses the "raster" (a square of 500×500 m in which information can be captured on our geographical information system, or GIS). MABA comprises 24 "partidos" of the Province of Buenos Aires which make up a continuous urban area with the Autonomous City of Buenos Aires.[5] The latter has not been included in the analysis as there are no available data that allow us to identify industrial hazardousness, and therefore environmental risk in the methodological terms that are explained below.

---

[3] ISO 14001 is a voluntary certified external environmental management norm issued by a supranational auditing organism.

[4] The INDEC (2003) uses the official name of Argentina's largest city, "Agglomerate Great Buenos Aires", which consists of CABA plus 29 neighbouring "partidos".

[5] The present work considers that the MABA is made up of 25 administrative units: La CABA-Ciudad Autónoma de Buenos Aires, and 24 partidos of the Province de Buenos Aires: Almirante Brown, Avellaneda, Berazategui, Esteban Echeverría, Ezeiza, Florencio Varela, General San Martín, Hurlingham, Ituzaingó, José C. Paz, La Matanza, Lanús, Lomas de Zamora, Malvinas Argentinas, Merlo, Moreno, Morón, Quilmes, San Fernando, San Isidro, San Miguel, Tigre, Tres de Febrero and Vicente López.

The method for calculating *industrial hazardousness* developed in Chapter 4 is based on the accumulation of the potential impact of each industrial generator and takes into account not only the hazardousness at the location of the industry, but also its influence depending on the distance from where the hazard is generated. It requires as initial data the geographical location of each particular industry (coordinates of latitude and longitude) and a parameter that allows us to associate each industry with a potential impact. Argentinean national censuses do not include these data and the information that exists in each province is not homogeneous, however. The municipality is the smallest geographical unit for which industrial data can be obtained for the whole territory from the same source. The 2001 census is the source used to compile the data of the number of industries per municipality and the average number of employees (calculated as the total number of employees per sector divided by the number of establishments per sector). The information used was obtained from the web page of the National Institute of Statistics and Censuses (Instituto Nacional de Estadísticas y Censos, INDEC).

For this reason, for the case of Argentina, it was necessary to develop an alternative methodology that allows calculation of the accumulated hazardousness of the industrial activities located in each of the municipalities. The methodology calculates the potential hazardousness of each industry using an algorithm that estimates factors of emission per industrial sector. The emissions factor of a firm is expressed as the weight of pollutant particles emitted annually per employee, using as initial data the type of industry (as different industrial categories are similar to one another in terms of production processes and technologies) and the number of employees per industry (which indicates the magnitude of the plant).[6]

Conventionally, the intensity of emission is measured as the volume of emissions of contaminant particles per employee. However, a real measure of intensity of emissions should also take into account the impact that the emissions have on the quality of the air. A better indicator of this impact is the total weight of contaminant particles released to the atmosphere, which can be obtained by calculating the pollution intensity of emissions (weight of contaminant particles in the unit volume of emissions). This intensity of pollution varies according to the type of industrial process and the size of the plant. Contrary to common sense, bigger plants release less pollution-intensive emissions. The greater the size of the plant, the less density of polluting particles per volume of emissions and consequently the less impact on the quality of the air per unit of volume of emissions. This is because larger factories have higher chimneys which retain more particles as they rise up the chimney, resulting in fewer contaminant particles per unit of gaseous volume released. On the other hand, the pollution generated by large companies is discharged at greater

---

[6] The algorithm produced by Dasgupta and Wheeler (2001) is based on mean real values of pollution emitted per province. However, these means come from aggregate data that do not take into account regional and local variations due to factors of regional/local governability. Consequently, on applying this algorithm to individual firms we assume the firm under analysis follows the behaviour of the average firm, thus we are estimating a potential rather than real hazardousness.

altitudes and is transported greater distances before being deposited,[7] while air pollution generated by smaller firms is precipitated in the immediate vicinity of the plant. As a result, small generators of pollution are more potentially hazardous than larger ones per unit of emissions for populations that are closer to the plant. The models of dispersion used by Dasgupta and Wheeler (2001) suggest that each unit of volume of emissions of a small plant increases the pollution of the surrounding air up to 14 times more than a unit of emission from a large plant. For this reason our model of calculation of potential hazard stresses the importance of incorporating the effect of emissions of small and medium-sized pollutant firms. The algorithm used in our model was developed in the World Bank by Dasgupta and Wheeler (2001), and it represents factors of contamination per industrial sector in the specific production conditions of each sector in the South American context. These contamination factors measure the intensity of annual emissions of contaminant particles per employee and they vary according to the industrial sector and the size of the firm: small (1–20 employees), medium (21–100 employees) and large (over 100 employees).

Table 6.1 presents the emission factors for each industrial sector per employee. The second column corresponds to the Standard Industrial Classification (SIC) used in the USA for each industry sector. The factors of emission are expressed in tonnes of contaminant particles emitted annually per employee.

Of the 512 administrative units considered, the National Institute of Statistics has available information on the number of industries for 509. The hazardousness of each industrial sector was calculated as the product of the number of firms in that industrial sector by the total number of employees in the sector by the emissions factor corresponding to the average number of employees per industry in the sector (Table 6.1). The final industrial hazardousness per municipality was calculated as the sum of potential emissions per industrial sector. This value was then transformed into a scale of 1–5 indicating values of rising hazardousness for later analysis combined with social vulnerability.

The ranges of accumulated hazardousness and distribution of hazardousness per administrative unit are summarised in Table 6.2.

In order to estimate *industrial hazardousness in the MABA* the methodology uses a mathematical function that calculates cumulative effects of specific sources in a given area and assumes that the hazard generated by an industry is not limited to the point of origin, but rather is spread over an area of influence until certain limits that are determined by the magnitude of the hazardousness of that industry. This means that the influence of an industry of given potential hazard extends to neighbouring areas, and there is an accumulation of the effects of several industries within an area of influence.

In each case, the application of said function requires knowledge of: a) a factor that ponders the magnitude of hazardousness of each industry and b) the radius of

---

[7] In addition, short chimneys retain very few "fine" particles (diameter of less than 10 μ), which have greater impact on morbidity and death rate (CBI 1998).

6 The Case of Argentina

Table 6.1 Annual emission factors per industrial sector (tonnes/employee)

Industrial sector	US SIC*	Small	Medium	Large
Basic foodstuffs	20	0.0256	0.0647	0.2072
Tobacco products	21	0	0	0.0056
Textiles	22	0.0494	0.0143	0.0233
Apparel	23	0.0063	0.0015	0.0005
Wood products	24	0.8924	0.0817	0.0919
Furniture	25	0.0078	0.0012	0.0009
Paper	26	0.0545	0.0475	0.0498
Printing	27	0.0009	0.0016	0.0004
Chemicals industry	28	0.0978	0.0652	0.2709
Oil refining	29	0.2099	0.0495	0.0561
Rubber and plastics	30	0.1201	0.0128	0.0101
Leather products	31	0.0279	0.0111	0.0105
Glass	32	0.0076	0.0225	0.0343
Metal products	34	0.0437	0.0165	0.0158
Computing and machinery	35	0.1319	0.0208	0.285
Electrical appliances	36	0.0138	0.0088	0.0501
Transport equipment	37	0.0093	0.0046	0.0116
Professional equipment	38	0.0309	0.0019	0.0006
Other manufacturers	39	0.0145	0.0026	0.0018
Beverages	208	0.4086	0.0264	0.0964
Other foodstuffs	209	0.1567	0.1792	0.0087
Other chemicals	289	0.018	0.0215	0.017
Oil products	299	0.2041	0.1111	0.2895
Footwear	314	0.0005	0.0003	
Ceramics	326	0.0172	0.0109	0.0029
Other non-metallics	3229	0.0565	0.1213	0.0046
Iron and steel	331–333	0.2112	0.235	0.0782
Non-ferrous	334–339	0.0433	0.0404	0.0972

Source: Dasgupta and Wheeler (2001)

Table 6.2 Index of industrial hazardousness (IP): ranges and frequencies

IP Qualitative ranges	Assigned value	Tonnes of annual pollution per department/partido	No. of departments/partidos (frequency)[a]
Very low	1	0–142	352
Low	2	143–415	98
Medium	3	416–799	43
High	4	800–1809	12
Very high	5	1,810–3,472	3

[a] The administrative political unit corresponding to Antarctic Argentina and the Islands of the South Atlantic is not included

influence of the hazard of that industry or R (distance to which the contaminant hazard might spread). The radius of influence of the potential hazard of an industry and its hazardousness are related by means of another mathematical function that increases the radius of influence depending on the proximity of other industries. In other words, the greater the hazardousness, the greater the radius of influence.

The indicator used to measure the magnitude of hazardousness in each industry was adopted based on the availability of accessible data. In the case of the MABA, the LEC or "Level of Environmental Complexity" is the chosen index for this purpose, as calculated by the Secretariat of Environmental Policy of the Province of Buenos Aires for each legally established industrial plant. The data used to calculate this index come from the information that the industry must present on declaring environmental impact. The LEC of each industrial establishment is defined by:

– The classification of each activity by industrial process (Ru), which includes the type of raw materials, of the materials handled, elaborated or stored, and the process carried out;
– The quality of the effluents and residues generated (ER);
– The potential risks of the activity (fire, explosion, chemical risk, acoustic risk) and risk due to machines working under pressure that may affect the surrounding population or environment (Ri);
– The size or magnitude of the plant, taking into account the number of employees, the energy installed and the surface area (Di).
– The location of the firm, bearing in the dominant use in the areas (residential, commercial, agricultural, recreational, etc) and the services infrastructure it possesses (Lo).

The LEC is expressed by means of a polynomial equation of five terms:

$$LEC = Ru + ER + Ri + Di + Lo$$

Depending on their LEC values, industries are classified the following categories:

– FIRST CATEGORY: values of up to 11
– SECOND CATEGORY: values of between 12 and 25
– THIRD CATEGORY: values of over 25

Establishments that elaborate and/or handle substances that are inflammable, corrosive, highly reactive, infectious, teratogenic, mutagenic, carcinogenic and/or radioactive, and/or that generate special residues are considered hazardous and are automatically classified as third category, irrespective of their LEC.

The procedures for calculating each of the components of the equation of LEC are explained in detail in Annexes I to VI of Decree No. 1345/1998, implemented through Resolution No. 177/2007 of the Secretariat for Sustainable Development (http://www.ambiente.gov.ar/?aplicacion=normativa&IdNorma=849&IdSeccion=214)

To calculate the value of R, or the distance over which the potential hazard might spread, a script in Avenue language was used, applying the procedure and formulae described in Chapter 4. As it was assumed that an increase in category corresponds to greater hazardousness, different values of R were taken depending on the category, and the distances of radiuses of influence used in the calculation were as detailed in Table 6.3.

Table 6.3 shows that industries in category 1 (NCA <11) are considered harmless by the legislation and therefore their radius of influence is zero. Industries in the second category (11<NCA<25) are small or medium generators with a radius of influence of 1,500 m. Finally, the radius of influence used for the generators of third category (NCA>25) is 3,000 m.[8]

As regards *social vulnerability*, the methodology of analysis is the same as in Chapter 3 of this book. As for hazardousness, the information used was taken from the National Census of Population 2001, disaggregate at departmental level, in absolute values (inhabitants or number of households).

The indicators used correspond to three key aspects or dimensions of social vulnerability (demographic, economic capacity and living standards), and they are reflected in Table 6.4.

These indicators were re-scaled to values of 1–5 using the technique of natural breaks to construct partial indicators of vulnerability. Finally the index of vulnerability was constructed as the simple addition of the values calculated for the 9 partial indicators in each administrative unit, once again applying the criterion of natural breaks offered by the Arcview programme.

The quantitative ranges express qualitative scales, which for the index of social vulnerability elaborated for Argentina corresponds to the values shown in Table 6.5.

The final calculation of risk was carried out by totalling the indices of IH and ISV and rescaling the values in ranges of 1–5 in line with the procedure described in Chapter 2.

The work goes on to describe and analyse the results obtained.

Table 6.3 Radius of influence of the hazardousness of an industry

Values of R adopted depending on the category	
Category	Radius of influence R
First	0
Second	1,500 m
Third	3,000 m

---

[8] According to the Law, there may be establishments that, despite having a LEC of less than 25, must be included in the third category due to the hazardous nature of their activity. In order to maintain the mathematical method of calculation and to avoid that a third category industry might be pondered with a lower coefficient than a second category one, for those industries a pondered value of LEC was adopted that was equal to the minimum for that category, i.e. 25.

Table 6.4 Social vulnerability in Argentina: indicators used

Dimensions	Variable	Indicator (corresponding to the number with which the basic table appears in the National Census of Population 2001)
Demographic	1. Transitory dependent population	1. Population in households by age group and gender, according to the type of household and the relationship with the head of family
	2. Definitive dependent population	2. Population in households by age group and gender, according to the type of household and the relationship with the head of family
	3. Single-parent homes	3. Head of household of incomplete nucleus, by civil status by gender and age group.
Economic capacity	4. Health cover	4. Population with health cover from a social charity and/or with a private or mutual health plan by gender and age group
	5. Literacy/education	5. Population of 10 or over by literacy and gender.
	6. Work/occupation	6. Population of 14 or over in employment or not economically active, by gender and age group.
Living standards	7. Basic needs	7. Households with more than 3 persons per bedroom
	8. Supply of drinking water	8. Households with the presence of this service in the segment
	9. Sewage services	9. Households with the presence of this service in the segment

Table 6.5 Index of Social Vulnerability (ISV): ranges and frequencies

ISV

Qualitative ranges	Assigned value	Sum values	No. of departments/partidos (frequency)[a]
Very low	1	9–13	354
Low	2	14–20	84
Medium	3	21–27	37
High	4	28–35	24
Very high	5	36–45	13

[a] The administrative political unit corresponding to Antarctic Argentina and the Islands of the South Atlantic is not included

4 Industrial Hazardousness in Argentina

Table 6.6 shows the hazardousness for each administrative unit in absolute terms (tonnes of annual contaminant emissions) and in relative ones (tonnes of pollution/km^2). The colour codes indicate the levels of hazardousness, from Very High to Very Low, corresponding to 5 levels of risk obtained using natural breaks in each variable. The column corresponding to the provinces has been coloured according to the final level of hazardousness of each province, obtained by averaging out the indices in columns 3, 4 and 5.

From Table 6.6 it can be seen that there is very high potential hazardousness in Buenos Aires, Córdoba, Mendoza and Santa Fe, and high hazardousness in Tucumán, Entre Ríos, San Luis and Capital Federal. These eight administrative districts represent a first, very general approximation to the diagnosis of hazardousness,

Table 6.6 Industrial emissions of contaminant particles in Argentina

Province	(1) Surface area (Km^2)	(2) Inhabitants	(3) Emissions (Tonnes)	(4) Emissions/inhabitant	(5) Emissions/Km^2
Buenos Aires	307,804	15,052,177	112,232	0.007	0.365
Capital	200	2,965,400	15,074	0.005	75.37
Catamarca	99,818	388,416	2,451	0.006	0.025
Chaco	99,633	1,052,185	3,833	0.004	0.038
Chubut	224,686	460,684	2,771	0.006	0.012
Cordoba	165,321	3,340,041	35,661	0.011	0.216
Corrientes	88,199	1,013,443	4,215	0.004	0.048
Entre Rios	78,781	1,255,787	10,894	0.009	0.138
Formosa	72,066	539,883	1,601	0.003	0.022
Jujuy	53,219	679,975	4,755	0.007	0.089
La Rioja	89,680	341,207	2,418	0.007	0.027
Mendoza	150,839	1,729,660	18,685	0.011	0.124
Misiones	29,801	1,077,987	5,131	0.005	0.172
Neuquen	94,078	547,742	1,831	0.003	0.019
RioNegro	203,013	597,476	3,751	0.006	0.018
Salta	154,775	1,224,022	6,828	0.006	0.044
San juan	89,651	695,640	5,044	0.007	0.056
San Luis	76,748	437,544	4,677	0.011	0.061
Santa Cruz	243,943	225,920	1,054	0.005	0.004
Santa Fe	133,007	3,242,551	35,010	0.011	0.263
Santiago	135,254	865,546	3,119	0.004	0.023
Tierra del Fuego	1,002,352	126,212	520	0.004	0.001
Tucuman	22,524	1,995,384	11,610	0.006	0.515

Colour Code	Range of Hazard	Index
	Very high	5
	High	4
	Medium	3
	Low	2
	Very low	1

as they represent 75% of the total population of Argentina and 35% of its surface area (excluding Argentinean Antarctic).

The map of hazardousness in administrative districts of Argentina (see Fig. 6.1) allows us to locate hazardousness in the territory with greater precision, absolute potential hazardousness or the accumulation of potential hazard in each political

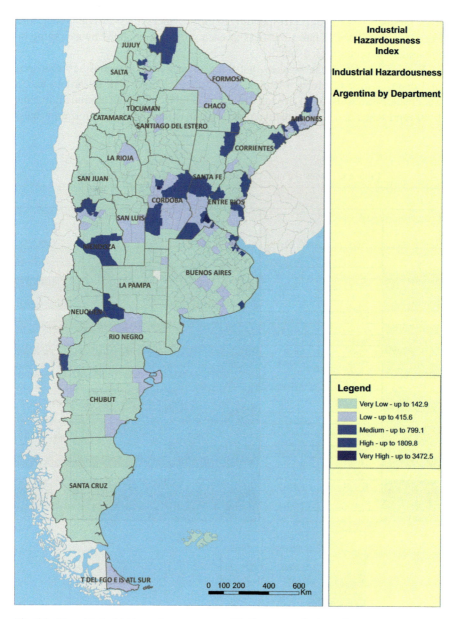

Fig. 6.1 Map of industrial hazardousness per partido/department in Argentina
Note: the darker the more hazardousness

unit. However, as the basic information does not allow us to locate the industry, the map does not identify those areas of the department with the highest density of pollution, since accumulated hazardousness is assigned to the whole of the administrative unit, and we are not able to distinguish whether it is concentrated into small, well-defined areas or spread over the whole territory.

Despite these limitations, the map in Fig. 6.1 provides information on industrial hazardousness at department level, classifying them into 5 categories: Very High Hazardousness, High Hazardousness, Medium Hazardousness, Low Hazardousness and Very Low Hazardousness. This information allows us to identify three groups of provinces, using as classification criterion the presence of administrative areas with high hazardousness in the territory (1 with high hazardousness, 2 with medium hazardousness and 3 with low hazardousness). The criterion used provides a more suitable indicator of hazardousness for the analysis of local risk than the aggregate values in Table 6.6. By analysing the spatial distribution of the departments/partidos affected, we can observe that the greatest hazardousness is located in a strip of 250 km along a diagonal joining the province of Misiones (in the northeast) and the southern tip of Neuquén (in the centre-west). To the south of this diagonal low hazardousness predominates, while to the north there is a medium level. Moreover, isolated situations of high hazardousness can be observed in the provinces of Salta and Jujuy.

High Hazardousness in Provinces: corresponds to those provinces that have at least one department or partido in a situation of very high or high hazardousness. The provinces included in this range are Provincia de Buenos Aires, Córdoba, Santa Fe, Entre Ríos, Tucumán, Mendoza, Misiones and Río Negro.

Medium Hazardousness in Provinces: corresponds to those provinces that have no department or partido in a situation of high or very high hazardousness but do have a) one or more departments or partidos in a situation of medium hazardousness; and/or b) fewer than 60% of departments or partidos in a situation of very low hazardousness. The provinces in this range are CABA, Neuquén, Chaco, Corrientes, Salta and Jujuy.

Low Hazardousness in Provinces: corresponds to the remaining provinces that have 60% or more of the departments or partidos in situations of very low hazardousness. The provinces in this range are San Juan, San Luis, Santa Cruz, Chubut, Catamarca, La Rioja, Tierra del Fuego and Santiago del Estero.

Comparing with the classification of provinces with high and very high hazardousness in Table 6.6, we can see that the spatial analysis in Fig. 6.1 has identified a situation of high hazardousness in Rio Negro and Misiones which was not reflected in results using the aggregate index presented in Table 6.6. The opposite occurs with San Luis and the CABA, which present medium-low hazardousness in the spatial analysis map although the aggregated index in Table 6.3 had classified them in the high hazardousness category. This indicates that although Rio Negro and Misiones have fewer total contaminant emissions than San Luis and the CABA, their emissions are located in a certain geographical cluster, while the emissions of the latter two are distributed over several administrative units.

Table 6.7 lists the 14 departments or partidos with high/very high Index of Industrial hazardousness/perilousness (IH).

Table 6.7 Departments with high and very high Index of Industrial Hazardousness (IH)

	Department or partido	Province	IH
1	Rosario	Santa Fe	5
2	Capital	Córdoba	5
3	General San Martín	Buenos Aires	5
4	Capital	Tucumán	4
5	General Roca	Río Negro	4
6	El Dorado	Misiones	4
7	Guaymallen	Mendoza	4
8	Federación	Entre Ríos	4
9	Vicente López	Buenos Aires	4
10	Tres de Febrero	Buenos Aires	4
11	La Matanza	Buenos Aires	4
12	Lanus	Buenos Aires	4
13	La Plata	Buenos Aires	4
14	General Pueyrredón	Buenos Aires	4

Note: IH: 5 = very high hazardousness; IH = 4 high hazardousness

The sum of the populations of these departments accounts for 15% of the Argentine population. The province of Buenos Aires has the highest number of political administrative units in situations of high or very high hazardousness (7 partidos, 3,110,000 inhabitants, 8% of the total population). With the exception of General Pueyrredón, located in the southeast of the province, all these partidos are located in the metropolitan area of Buenos Aires (MABA), which has been taken as a case study and will be analysed more thoroughly in the following section.

5 Social Vulnerability in Argentina

Figure 6.2 shows the social vulnerability in Argentina distributed by departments or partidos, for each province and for the country as a whole. The worst situations can be observed in the centre-north of the country, with a high number of departments with major cities, including provincial capitals.

Within this general scenario in the centre-north of the country there are extreme situations. On the one hand the province of La Pampa has a low or very low profile of social vulnerability, and in Catamarca, La Rioja and San Juan low values predominate, with the exception of those departments in which the provincial capitals are located, in which case the values are in the medium range.

On the other hand, there are situations of higher social vulnerability, as the following Table 6.8 illustrates.

The values of very high social vulnerability (range 5 in our classification) correspond to 7 partidos of the Province of Buenos Aires, which together with the CABA are part of the MABA; the partidos of La Plata (provincial capital) and of Gral. Pueyrredón (home to Mar del Plata, the province's second city in terms of population); the department of Rosario, which is home to the largest city in the

6 The Case of Argentina

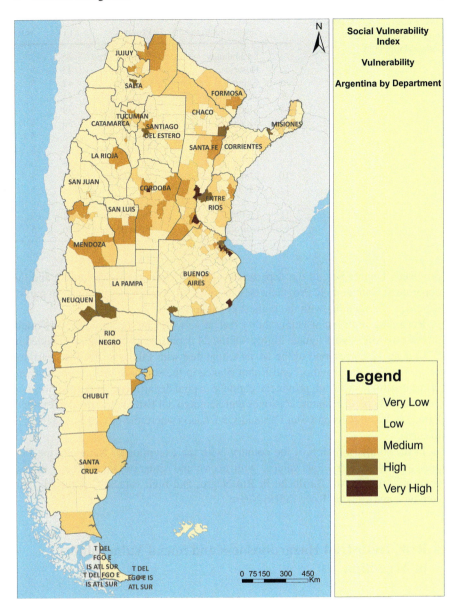

Fig. 6.2 Map of social vulnerability in Argentina by department/partido
Note: The darker the more social vulnerability

Table 6.8 Departments with higher Index of social vulnerability (ISV)

	Department/partido	Province	ISV	Value
1	La Matanza	Buenos Aires	45	5
2	Rosario	Santa Fe	43	5
3	Capital	Córdoba	42	5
4	Almirante Brown	Buenos Aires	41	5
5	Merlo	Buenos Aires	39	5
6	Moreno	Buenos Aires	39	5
7	Florencio Varela	Buenos Aires	38	5
8	General Pueyrredón	Buenos Aires	38	5
9	La Capital	Santa Fe	38	5
10	Lomas de Zamora	Buenos Aires	38	5
11	Ciudad Autónoma de Buenos Aires	–	37	5
12	La Plata	Buenos Aires	37	5
13	Quilmas	Buenos Aires	36	5

province of Santa Fe and the department of Santa Fe, home to its capital; and the department Capital de la Provincia de Córdoba. All of them include or form part of the cities of Argentina with the highest populations.[9]

Regarding high social vulnerability (range 4 of the classification) the Provincia de Buenos Aires again figures highly. Of the 24 departments or partidos included in this range, 12 correspond to the MABA, together with Pilar and Escobar, partidos in the north of the province, and Bahía Blanca in the south. In this group there are also departments who are provincial capitals: capital departments in the provinces of Tucumán, Salta, Corrientes, Misiones and Santiago del Estero; and the departments Paraná in Entre Ríos and San Fernando in Chaco (where the city of Resistencia is located).

In the southern region of the country, with the exception of Neuquén, the ISV presents low or very low values, with occasional situations of medium values: Rawson[10] in Chubut, Bariloche in Río Negro; and high values: Confluencia[11] in Neuquén and General Roca in Río Negro.

6 Risk, Industrial Hazardousness and Social Vulnerability in Argentina

The map of industrial risk (Fig. 6.3) shows the application of the methodology proposed in this book as an approximation that allows us to carry out diagnoses that are better suited to identifying areas of industrial risk where intervention is a priority. As

---

[9] This configuration is the result of the weight of working with absolute values.

[10] Home to the provincial capital.

[11] Home to the provincial capital, this city has a functional continuity with the city of Cipolletti, in Gral. Roca, Province of Río Negro.

6 The Case of Argentina

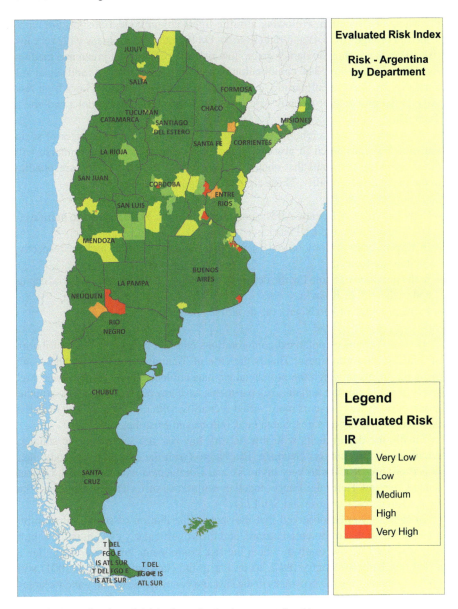

Fig. 6.3 Map of evaluated risk in Argentina by departments/partidos
Note: The darker the higher risk

explained in the chapter on the methodology and theoretical framework, the result of combining the indices of industrial hazardousness (IP) and social vulnerability (ISV) enables us to identify the so-called "hot spots", i.e. areas with high (4) or very high (5) combined risk (IR). Figure 6.3 presents the map of evaluated risk for the whole of Argentina. Following a traffic-lights colour code, hot-spots are represented in orange (IR = 4) and red (IR = 5). On the other hand, areas with very low risk (IR = 1) are show in darker green, low risk (IR = 2) in light green and medium risk (IR = 3) in yellow. The combination of industrial hazardousness and social vulnerability by department/partido show eight geographical areas of very high risk, three in the province of Buenos Aires, two in Santa Fe, one in Córdoba and one in Tucumán.

As a result of the weighting given to the population for each indicator of social vulnerability considered in absolute values, the eight areas of risk coincide with those administrative units that obtain the maximum ISV value of 5, corresponding to the highest vulnerability. These eight areas are the following:

– In the province of Buenos Aires: some partidos of the MABA, and the partidos of Gral. Pueyrredón and La Plata.
– In the provinces of Córdoba and Tucumán, the departments in which the respective capitals are located.
– In the province of Santa Fe: the departments that are home to the provincial capital and to the city of Rosario.
– In the province of Río Negro, the department Gral. Roca.

In some provinces polarised configurations can be observed in the distribution of industrial hazardousness. This is the case of Tucumán and Chaco, where only the departments in which the provincial capitals are located have high or very high levels of risk, while the remaining departments have low or very low values. Another group of provinces presents more varied scenarios, with different degrees of risk in their territory: Buenos Aires, Santa Fe, Córdoba and Misiones.

Finally the departments with low or very low risk are located in the southern provinces of the country: Neuquén, Río Negro (with the exceptions already mentioned of Gral. Roca with very high risk and Confluencia with high risk), Chubut, Santa Cruz and Tierra del Fuego, together with those of La Pampa, La Rioja, San Juan, Catamarca, Formosa and San Luis.

In short, it can be observed that the spatial distribution of social vulnerability coincides with that of potential hazard, and this is particularly evident in the central diagonal band, to the south of which the risk components are lower. Argentina is a highly urbanised country (89% according to the National Census of 2001) with higher population in the provinces in the east and centre of the country and its largest cities are located along this diagonal.[12] This observation is in line with empirical

---

[12] A natural continuation of this research would be to include a second period of processing and analysis of the social information, not considering the values of each indicator in absolute terms (number of inhabitants, number of households) as in the present study, but rather in relative terms, considering the percentage of total inhabitants/households that make up each of the five ranges contemplated. A combination of the two approaches would allow us to identify the situations of greatest vulnerability: those in which the values are very high or high in both absolute and relative terms.

research carried out in the USA, which, using similar geographical units of analysis, found evidence of environmental injustice in the distribution of urban waste tips (US GAO 1983; Heitgerd et al. 1995), environmental contamination (Liu 2001; Perlin et al. 1995) and environmental hazardousness in general (Been and Gupta 1997).

Based on maps such as this one it is feasible to carry out research in greater detail, choosing case studies that are most relevant for the analysis of industrial risk on the basis of real data and explicit variables. On this occasion the MABA has been chosen as a case study for the application of the methodology on a more detailed scale. The analysis of the results obtained is outlined below.

7 Analysis of the Administrative Units of the MABA

7.1 Industrial Hazardousness (IP) in the Metropolitan Area of Buenos Aires

The maps drawn up with the information referring to industrial hazardousness show the concentration of industries in the metropolitan area of Buenos Aires (Fig. 6.4) and the accumulated hazardousness (Fig. 6.5) expressed by the colour code, where the greatest degree of hazardousness corresponds to the darkest tone of blue.

The map showing the concentration of industries reveals two different areas of great industrial activity to the northeast and southeast. These areas are clearly separated by the basin of the Matanza River to the southwest. The area to the northeast is located between the Reconquista River and the Federal Capital, while the one to the southeast corresponds to the provincial jurisdiction of the Matanza-Riachuelo basin. The industrial density is greater in the former area, which is organised as a consolidated belt around the northern and western limits of the Federal Capital, with a structure of "short fingers" with decreasing intensity from the perimeter of the Federal Capital towards the interior of the province in four departments/partidos with high industrial concentration: Vicente López, San Martín, Tres de Febrero and La Matanza. In the southeastern zone the industrial concentration has a clear core in La Matanza and Lanús, which then spreads along two axes, one to the southeast and the other, of less density, to the south.

The analysis of industrial hazardousness based on this map (in which the industrial concentration is used as an indicator of potential hazard) identifies six municipalities with greatest hazardousness: Vicente López, San Martín, Tres de Febrero, La Matanza, Avellaneda and Lanús. In all cases, the greatest potential risk is accumulated in a stretch of 6 km from the Federal Capital. The consequences of this distribution of governability of industrial hazard may be positive, as the concentration of generators in areas with a good communications infrastructure would allow the necessary governmental resources to be organised more efficiently by means of inspection campaigns.

The map of hazardousness in Fig. 6.5 allows us to diagnose with greater accuracy which areas have the greatest potential hazard. Although the partidos mentioned in

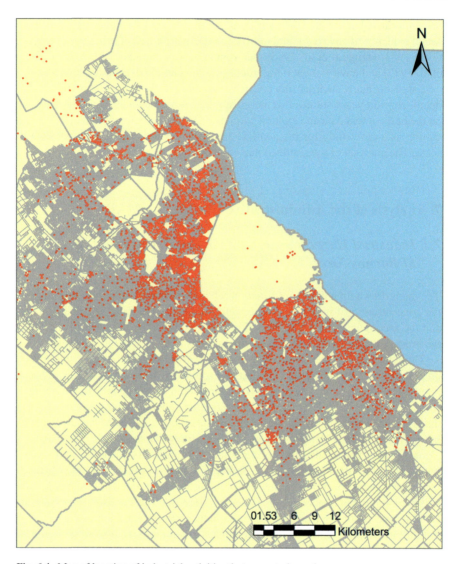

Fig. 6.4 Map of location of industrial activities that generate hazard

the previous paragraph can be identified as areas of high hazardousness, the maximum hazard is focused around two nodes. The first one is located to the west of the municipality of Vicente López and has an area of influence of one kilometre which stretches as far as the municipality of General San Martín; the second is completely located in the southern zone of the municipality of General San Martín.

The map also shows small areas of high risk in municipalities that are further away from the industrial belt which borders the Federal Capital: Tigre, Malvinas Argentinas, Hurlingham, Morón, Quilmes, Esteban Echeverria and Almirante

6 The Case of Argentina

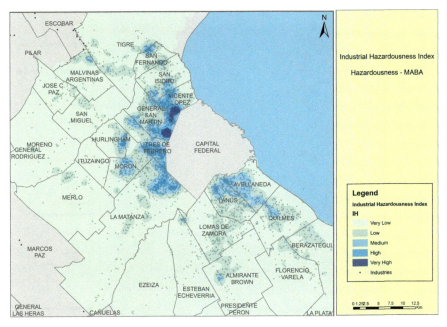

Fig. 6.5 Map of industrial hazardousness in the metropolitan area of Buenos Aires
Note: The darker the more hazardousness

Brown. The special analysis based on this map shows the potential hazard arranged as a triangle whose base is on the Río de la Plata, with the greatest hazardousness concentrated in the northeastern sector. In addition to this sector of "continuous hazard", there are "isles of hazardousness" linked by areas of low risk surrounding the lines of communication. In terms of governability, this implies that the scenario for increasing the effectiveness of controls is more complex, and as a result it is more difficult to specify.

7.2 Social Vulnerability in the Metropolitan Area of Buenos Aires

Each of the 24 "partidos" of the province of Buenos Aires making up the MABA presents social characteristics of great heterogeneity that the data aggregated per partido do not reflect. Bearing this in mind, and given the fact that the information on industrial hazardousness is provided for cells of 0.5×0.5 km. (while the partidos considered have surface areas of between 36 and 924 km^2), indicators of vulnerability were sought on the scale of the "radiuses or block groups", the smallest units of census data dissemination available, which bear a certain resemblance to the above-mentioned cells. The block groups of the 24 partidos of the MABA for the census of 2001 totalled 8,011. Regarding the indicators selected to construct the ISV, the same ones were used as for the national study (see Table 6.4).

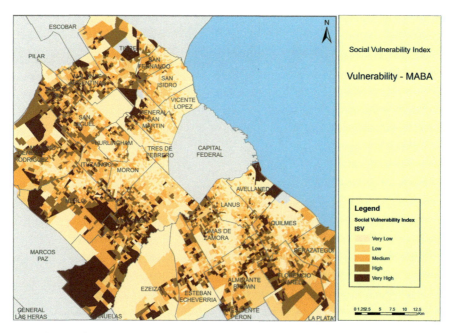

Fig. 6.6 Map of social vulnerability in the metropolitan area of Buenos Aires by census unit (block groups)
Note: The darker the more social vulnerability

Unlike the analysis carried out for the whole country, the map of social vulnerability by block group in the MABA presented in Fig. 6.6 shows clear differences with the spatial distribution of industrial hazardousness. A first belt can be observed around the CABA in which low social vulnerability predominates with the exception of the partidos located in the south, with several block groups with high and very high social vulnerability: the coastal sector of Avellaneda, home to the petrochemicals industry in Dock Sur, Lanús and Lomas de Zamora.

A second belt further away from the Federal Capital includes several areas of high and very high social vulnerability: to the south in Almirante Brown and Florencia Varela; to the southwest in the partidos of La Matanza, Merlo and Moreno; to the northeast in José C. Paz, Malvinas Argentinas and San Miguel.; and to the north a north-south vector joining the north of General San Martín, passing through the west of San Fernando and continuing to the centre-north of Tigre.

On the whole, on this scale of analysis, the trend of spatial distribution seems to contradict the hypothesis of "environmental injustice" which establishes a correlation between hazardousness and vulnerability. Although research in situ and statistical analysis are required to confirm the tendencies observed, the spatial distribution of the map suggests an inverse relationship between vulnerability and hazardousness, in which the gradient of hazard decreases at greater distances from the CABA while the gradient of social vulnerability increases. The difference between these conclusions and those obtained in the previous section using another scale of analysis go to confirm the warnings of Bowen (2002) regarding the validity of studies on environmental injustice using geographical units of analysis that do not allow examination of local changes in the distribution of tendencies of risk. However, we should bear in mind some additional considerations regarding these differences. On the one hand, in the national case the geographical basis of aggregation of the data used in the calculations of both industrial hazardousness and social vulnerability is the same: information on the level of departments/partidos provided by the National Census of Population, Family and Housing of 2001. However, for the MABA the units of analysis of both components of risk differ at least in their geographical definition, which does not come from the same source (industry information comes from Buenos Aires Province Direction of Statistics). On the other hand, the ISV on this scale reveals, in general terms, the expulsion of the most vulnerable population towards the less consolidated peripheral areas, where land prices are lower and where there is no guarantee of drinking water, sewage or public transport.

In this sense, it proves interesting to observe and discuss the map that is obtained from the combination of these components in the Index of Evaluated Risk.

7.3 Evaluated Risk in the Metropolitan Area of Buenos Aires

Figure 6.7 presents the map of industrial risk of the metropolitan area on the level of block groups or census radius. The map shows that the situations of high and very high risk (arising from the combination of very high and high values of both industrial hazardousness and social vulnerability) are focused on areas that are scattered around the whole conurbation, in most cases forming isles of high and very high risk in the midst of areas characterised by low levels of industrial risk. Exceptions to this overall panorama are the partidos of La Matanza and to a lesser degree Gral. San Martín, Malvinas Argentinas and Quilmes; where the areas of high and very high risk are surrounded by areas of moderate and medium risk.

To obtain a quantitative assessment of the magnitude of population at risk, we used GIS tools to calculate the intersection between the spatial the Layer containing population data and the polygons defining areas of high and very risk. The results of the analysis show that in total, 108.344 inhabitants of the metropolitan area of Buenos Aires live under high or very high environmental risk.

Both the patterns of Industrial location and the settlements of vulnerable people tends to follow the presence of radial lines of communication that connect the periphery and the provinces with the centre of the conurbation, the Autonomous City of Buenos Aires. Such confluence of allocation of economic and residential activities around particular roadways may be associated and may explain to a certain degree the presence of linear strips of territory with high and very high values of risk: one runs along the southeast of Tigre, through the centre of San Fernando

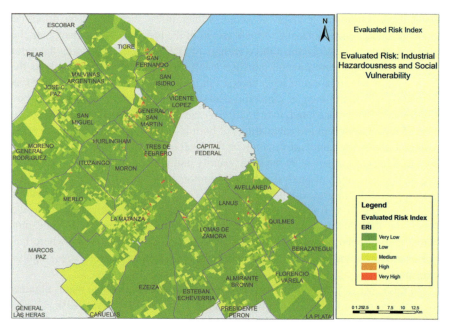

Fig. 6.7 Map of evaluated risk in the metropolitan area of Buenos Aires by census unit (block group)

ending in the north of Vicente López, and this coincides with a branch of the so-called Pan-American motorway; another crosses the partido of La Matanza from northeast to southeast and coincides with a national highway leading to the south of the country (RN3).

Finally, two clusters of block groups under medium-to-high risk can be observed on the map in the shape of a cloud or spot of certain magnitude and continuity including values of medium, high and very high industrial risk: one located in the northeast of the partidos of General San Martín and Tres de Febrero, and another in the south of Avellaneda and Lanús, continuing to the east of Quilmes.

8 Conclusions and Future Works

It should be stressed once more that this work is a starting point to analyse the topics in question in a transparent manner. As the assumptions under which the work has been carried out have been made clear, the results can be reviewed from many points of view by modifying the techniques applied explicitly and experimentally.

In this sense, the work presented in relation to social vulnerability only refers to values of synthesis and it did not include the analysis of each of the indicators considered. It would also be enlightening to consider the indicators of SV in relative terms, i.e. taking into account the number of inhabitants/households in unfavourable conditions as a proportion of the totals in each political administrative unit. Finally, in addition to these results in absolute and relative terms, the characterisation would be more complete and interesting if sub-indices were elaborated and analysed grouping together the indicators in consistent thematic dimensions: demography, economic capacity and living standards.

In terms of hazardousness, the methodology used also has its limitations. The criterion of using the best values available for each scale of intervention leads to a better identification of areas of potential hazard to pinpoint interventions, but at the same time it reduces the comparability of data among different scales. It would therefore be of interest to contrast the conclusions obtained in the MABA with alternative maps using the algorithms of the World Bank. In addition, in order to be able to generalise its application in diverse contexts, the methodology and algorithms used to accumulate the individual impacts in a given area should be improved by incorporating geographical variables such as the effects of relief in the diffusion of contamination. These effects were not included in the design as that would have implied a substantial increase in the cost of acquiring data, which would not have been compensated by the limited aggregate value to be obtained in the case study of the MABA, an area of predominantly flat land in which the impact of the terrain on accumulated hazardousness is marginal.

Calculation of the index of risk supposes that exposure can be calculated as the arithmetic sum of the hazardousness and vulnerability vectors in a geographical unit of analysis. However, the negative synergies between hazardousness and vulnerability that create "vicious circles" of risk might be calculated better using other more complex mathematical functions. Similarly, the final result might be different if, rather than adding the indices ISV and IP, the risk had been obtained by totalling the variables used to create the indices and then determining the ranges of risk using natural breaks. Finally, the quality of the existing data implies that the maps obtained reflect a historical situation rather than a true diagnosis of the current situation. The census data used for calculating vulnerability correspond to the census of 2001 (the only one available in 2008 when this project commenced), and so for the sake of consistency the algorithms for calculating industrial hazardousness and the databases of industries are from the same year. In the decade that has since passed, significant changes may have taken place regarding infrastructure, demographic distribution and location of industries. Also new technologies may have improved the efficiency of the industrial processes.

The above-mentioned limitations influence the dimension of "uncertainty" described in our theoretical framework in Chapter 3 and imply that the results obtained must be considered with caution. This is not a diagnosis of current risk, but rather an exercise in applying a conceptually innovative methodology with historical data, providing a "snapshot" of risk in 2001. The natural continuation of this study would be fieldwork in the areas identified, refining procedures and comparing several methodologies of analysis with samples of air/water quality and quantitative and qualitative analyses of evolution of the indicators of vulnerability and hazardousness over the last decade.

References

Been, V., & Gupta, F. (1997). Coming to the nuisance or going to the barrios? A longitudinal analysis of environmental justice claims. *Ecology Law Quarterly,* 24, 1–56.

Bowen, W. (2002). An analytical review of environmental justice research: What do we really know? *Environmental Management, 29*(1), 3–15.

Chudnovsky, D., Pupato, G., & Gutman, V. (2005). *Environmental management and innovation in argentine industry. Determinants and Policy Implications.* Documento Tecnico N° 36 (Fundación CENIT, Buenos Aires).

CBI. (1998). *Worth the Risk: Improving Environmental Regulation.* London: Confederation of British Industry.

CIA. (2011). *The World FactBook.* Central Intelligence Agency, Washington, USA. Retrieved September 7, 2011, from https://www.cia.gov/library/publications/the-world-factbook/geos/ar.html

Dasgupta, S., & Wheeler, D. (2001). Small plants, industrial pollution and poverty. In R. Hillary (Ed.), *Small and medium-sized firms and the environment* (pp. 289–304). Sheffield: Greenleaf Publishing.

ECLAC. (2010). *Statistical yearbook for Latin America and the Caribbean.* Economic Commission for Latin America and the Caribbean.

Heitgerd, J. L, Burg, J. R., & Strictland, H. G. (1995). A geographic information systems approach to estimating and assessing national priorities list site demographics. Racial and Hispanic origin composition. *International Journal of Occupational Medicine and Toxicology, 4*(3), 343–363.

Hochstetler, K. (2002). After the boomerang. Environmental Movement and Politics in the La Plata River Basin. *Global Environment Politics, 2,* 35–58.

Hochstetler, K. (2003). Fading green? Environmental Politics in the Mercosur Free Trade Agreement. *Latin American Politics and Society, 45*(4), 1–32.

INDEC. Instituto Nacional de Estadística y Censos. (2003). ¿Qué es el Gran Buenos Aires? 12 p. Versión digital en. www.indec.gov.ar

Liu, F. (2001). *Environmental justice analysis. Theories, methods and practice.* Boca Raton: Lewis Publishers.

Newell, P., & Muro, A. (2006). Corporate social and environmental responsibility in Argentina: The evolution of Agenda. *Journal of Corporate Citizenship, 24,* 49–68.

Perlin, S. A., Seltzer, W., Creason, J., & Sexton, K. (1995). Distribution of industrial air emissions by income and race in the United States: An approach using the toxic release inventory. *Environmental Science and Technology, 29*(1), 69–80.

US GAO. (1983). *Siting Hazardous Waste Landfills and their Correlation with Racial Status of Surrounding Communities.* United States General Accounting Office.

Vazquez-Brust, D., Plaza-Úbeda, J. A., Natenzon, C., & Burgos-Jimenez, J. (2009). 'The challenges of businesses' intervention in areas with high poverty and environmental deterioration: Promoting an integrated stakeholders' approach in management education'. In C. Wankel & J. Stoner (Eds.), *Management education for global sustainability* (pp. 175–206). New York: Information Age Publishing.

Vazquez-Brust, D., Liston-Heyes, C., Plaza-Úbeda, J., & Burgos-Jimenez, J. (2010). CSR, stakeholders' management and stakeholders integration in Latin-America. *Journal of Business Ethics, 91*(2), 171–192.

Vives, A. (2006). Social and environmental responsibility in small and medium enterprises in Latin America. *Journal of Corporate Citizenship, 21,* 39–50.

Chapter 7
The Case of Spain

José A. Plaza-Úbeda, Julieta Barrenechea, Jerónimo de Burgos-Jiménez, Miguel Pérez-Valls, and Sergio D. López

Abstract This chapter presents the evaluation of environmental industrial risk in Spain following the methodology outlined in previous chapters. The empirical data collected allowed us to calculate a risk index for the whole of Spain and to develop more detailed spatial analysis for two specific case studies, the cities of Madrid and Seville. The risk index at the national level identifies the Spanish towns at greatest risk from the combined factors of social vulnerability and industrial hazardousness, while the case studies' findings show that there is no "hot-spot" in Madrid but certain areas of Seville are exposed to very high combined risk.

Keywords Environmental risks · Industrial hazardousness · Spain · Madrid · Seville

1 Background: The Spanish Context

Spain is made up of 17 Autonomous Communities and 2 Autonomous Cities and it takes up most of the Iberian Peninsula in the southwest corner of Europe. The Spanish territory is characterised by a great geographical, climatological and biological diversity (with examples of four of Europe's 10 bio-geographical regions), which together with social, cultural, political and economic aspects make up a territory that is both plural and unique. Spain's approximate 10,099 km of coastline account for almost 15% of the total coastline of the European Union (EU).

Spain has undergone rapid demographic growth in recent years, contrary to forecasts made in the 1990s which did not foresee a population of over 40 million in the short term. Indeed, in 2008 Spain's population reached 46,157,822 inhabitants (Ministerio Medio Ambiente 2008), and over the period 1996–2008 there was a 16.4% rise in population. However, the analysis of autonomous communities shows considerable differences in population growth. For example, while the Balearic Islands experienced 41% growth, the population of Asturias fell by 0.7%, with high increases in the autonomous communities along the east coast, the Canary Islands and Madrid (Ministerio de Medio Ambiente 2008). This population rise was

J.A. Plaza-Úbeda (✉)
Department of Business Administration, University of Almeria,
Almeria 04120, Spain
e-mail: japlaza@ual.es

due both to net migration[1] (which is also chiefly responsible for the rise in birth rate) and to natural growth of the population.

In terms of population, Spain ranks fifth among the 27 EU members, with 9% of the total population, behind Germany (16%), France (13%), the United Kingdom (12%) and Italy (12%) (Ministerio de Medio Ambiente 2008). However, according to data from 2008, the density of population in Spain, 91.2 inhabitants/km^2, was one of the lowest in the EU (only 8 countries had a lower density).

From the economic point of view, net available national income per capita at market prices at the end of the 1990s was around 17,000 Euros, whereas at present it is over 26,000 Euros. This represents an increase of over 50% since the turn of the century (INE 2005, 2009). Nevertheless, the current economic crisis has had an alarming effect on the economy and the population, and the unemployment rate is one of the main indicators of this scenario.

Spanish economy is traditionally influenced by a high dependency of three main sectors: agriculture, tourism and building. Most of the industries are based in the North of the country, whereas the South sustains tourism and agriculture.

The North (the Basque country) holds the majority of heavy industry (iron and steel); in the Northeast (Catalonia) sustains chemical industries and those manufacturing by-chemical products (plastics and others). The Centre (Community of Madrid) and East (Community of Valencia) develop a wide range of manufacturing industries, together with Catalonia. Heavy naval industry is distributed along our extended coastline, in Bilbao (North), El Ferrol (Northwest) and Cadis (South-Andalusia). Likewise, the oil industry can be located in Tarragona (Catalonia), Cadis and Huelva (South of Andalusia). The automotive industry is also important, in Zaragoza, Barcelona and Andalusia, places where a large number of multinational companies are established.

In Spain, water, soil and air pollution by the industry is a reality mainly motivated by economic development. According to Greenpeace (2008), the presence of pollutants is not sufficiently documented in Spain. But the existent information allows for a high level of concern. The abovementioned Greenpeace report analyses the location of chemical and concrete Spanish industries and points at Andalusia, Catalonia and the Basque Country as the Autonomous Communities with the highest levels of hazardousness.

The main sources of chemical pollution are industrial emissions and spillage, waste handling and hydrocarbons. According to the European Environment Agency, industrial production and commerce contribute in 41.4% to soil pollution; spillage and urban waste handling in 15.2%; and oil industries in 14.1%.

In the field of business, environmental issues are of ever greater concern and over recent years a set of policies have helped to ensure that corporate practices are more and more environmentally responsible. The main tendencies observed in Spanish

---

[1] According to the National Survey of Immigrants (INE 2008), in 2007 the number of immigrants in Spain was over four and a half million, accounting for over 10% of the country's total population.

business have been directed towards increasing expenditure on protecting the environment, increasing environmental regulations and environmental quality standards and motivating greater management concern for environmental issues. According to data from the Ministerio de Medio Ambiente (2008), over the years 2000–2007 an improvement was observed in Spain in terms of GDP regarding consumption of non-renewable energies (coal, oil, and nuclear energy), consumption of petrol and fuel-oils, and emissions of SO_2, CH_4 and CO. However, the opposite occurred regarding emissions of greenhouse gases, CO_2, N_2O, NH_3 and consumption of diesel and kerosene.

There is abundant legislation related to the environment and nature protection on the European, national, autonomous government and in some cases even local government levels. Since joining the EU in 1986, Spain has undergone an intense process of adaptation in different socioeconomic aspects in order that its socioeconomic conditions and in particular national legislation fall in line with those of other member countries. This process has included major changes for firms in areas such as safety at work and environmental protection. At present, almost all the European institutional pressure regarding health care, safety and the environment has been taken on board by Spain. The EU makes a great effort to guarantee the physical welfare of its citizens by intensifying its activities in several fields that involve risks for the population: environmental contamination, radiation, noise, electromagnetic fields, improving product safety, industrial safety and safety at work. These actions are intended to reduce the risk to the population acting on two fronts: uncertainty and governability.

On the one hand a major effort has been made to improve the information on these possible risk scenarios, acting on uncertainty by providing easy access to the best scientific information available and publishing environmental information. Along these lines, Law 27/2006 on the access to environmental information has allowed the advent/development of Environmental Information Networks (in many cases with the competences transferred to the autonomous communities). These networks fulfil a triple role in providing technical information for managers and researchers, public information for citizens and indicators for the allocation of resources and decision making. In addition, several scientific committees have been set up on both EU and Spanish level with the dual aim of examining the toxicity and ecotoxicity of chemical, biochemical and biological compounds whose use might harm human health and the environment, and of advising on the elaboration and application of regulations. Also greater publicity is given to the environmental impacts of firms, for instance by publishing the levels of certain emissions in the EPER database.

On the other hand, the EU has taken pains to develop a legal framework that helps to guarantee a high level of safety for its citizens. This framework has involved coordinating national measures and establishing specific legislation in the form of Directives and Norms. In Spain, apart from EU norms (e.g. Directive 2008/1 of January 15th on Integrated Pollution Prevention and Control), there are environmental norms that condition the environmental impact of firms on the basis of "*he who pollutes, pays*". One of the most noteworthy of these norms is the Law of Environmental Responsibilities (Ley de Responsabilidad Medioambiental, Ley

26/2007), which stipulates that the operators of certain economic and professional activities must provide a financial guarantee against any possible future environmental damage that they might cause.

Other norms deal with more specific issues, for instance Real Decreto 9/2005 Suelos Contaminados (Contaminated Soil), Ley 10/98 de Residuos (Waste) – soon to be revised –, etc. Other norms that have had repercussions on risk are those of industrial safety, such as Ley 31/1995 Prevención de Riesgos Laborales (work related risks), Real Decreto 1254/1999 on the control of risks inherent in serious accidents involving dangerous substances, and their respective regulations.

All of the above is intended to show that although it is extremely useful to evaluate risk according to the methodology described in the present work, hitherto the effects of governability and uncertainty have not been considered, and this may mean that the gap between evaluated risk and managed risk is greater.

2 Data Collection and Methodology of Analysis

The mapping of potential risk due to environmental hazardousness and social vulnerability in Spain has been facilitated by the greater availability of information in comparison with Bolivia and Argentina. This has allowed the mapping of hazardousness and social vulnerability on a municipal level for the whole country and on the level of census units for the two cities analysed as case studies: Seville and Madrid.

The systemisation of information to calculate hazardousness (IH) began with the detection and positioning in GIS of the Spanish firms available for the indicators shown in the section of hazardousness due to contamination and the number of employees in each of the firms. Firms were selected from the "SABI" database with data from late 2008.

In the case of social vulnerability, systemisation of information was oriented to identify the data corresponding to the factors and indicators defined in the methodology of analysis of social vulnerability expounded in Chapter 3, taking into account demographic and economic dimensions as well as standard of living according to the data collected in the last National Census of 2001.

This chapter presents the results of the research for the case of Spain. It explains the methodology and analytical results for both environmental hazardousness and social vulnerability. It then goes on to provide the results of the analysis of combined risk, concluding with the analysis carried out for the specific case studies of the cities of Madrid and Seville.

3 Industrial Hazardousness in Spain

As was explained in Chapter 4, our proposal for calculating an index of industrial hazardousness is based on the accumulation of the potential impact of each industrial stressor for a large number of productive sectors. It takes into account not

7 The Case of Spain

only the hazardousness at the site of the industry, but also its spatial influence as a function of distance from the stressor. It requires the geographical location of each industry (coordinates) and a parameter that allows us to associate each industry with a potential impact. In order to assess the potential hazardousness for the whole of Spain, knowledge of the coordinates is necessary to be able to locate all the industries in potentially contaminating sectors. As explained in the methodology for the calculation of risk due to industrial hazardousness, the industries were classified in 3 categories according to their environmental impact depending on the industrial activity of each one.

Once this analysis had been carried out on a municipal scale for the whole of Spain, the methodology adopted provides a descriptive map of the different indices of hazardousness (see Fig. 7.1).

The map below provides a more complete view of this analysis, since the effect of hazardousness is represented on a municipal scale. The dark blue areas represent the focal points or places where hazardousness is greatest in Spain. It can be seen that *Barcelona and Saragossa are the only Spanish cities with a very high level of industrial hazardousness.*

Table 7.1 shows municipalities and resident population according to levels of Industrial Hazardousness. The breakdown of the information reveals that over 99% of Spanish municipalities (68.52% of the population) currently present low (2) or very low (1) levels of industrial hazardousness, whereas only 13 municipalities (14.01% of the population) present high (4) or very high (5) levels.

Fig. 7.1 Map of environmental hazardousness in Spain by municipalities
Note: the darker the more hazardousness

Table 7.1 Municipalities and resident population according to levels of IP

Levels of IP	No. of municipalities	% population
1	7,715	43
2	315	25
3	65	17
4	11	7
5	2	6
Total	8,108	100.00

To complete this approach we have selected the 13 municipalities with the highest index of industrial hazardousness ("very high" (5) or "high" (4)) with a view to making an initial assessment of the population exposed to this potential hazardousness. Table 7.2 shows all these municipalities, which include several provincial capitals: Barcelona, Saragossa, Malaga, Murcia, Seville, Valencia and Valladolid.

Generally speaking, the data of the analysis for the 8,098 Spanish municipalities show that 14.01% of the population (5,809,617 inhabitants) is exposed to high or very high levels of hazardousness and that this population is concentrated in 13 municipalities. Likewise, the data reveal that in some cases this hazardousness can be associated with higher concentrations of industrial zones, as in Barcelona, Saragossa and Seville, while in others this correlation is not so clear a priori (Malaga and Murcia).

The results of the analysis of industrial hazardousness are similar to previous studies carried out in the Spanish context. For instance, the 2008 Greenpeace report analyses in the main the location of chemicals and cement industries in each autonomous community, finding that Andalusia, Catalonia and the Basque Country have the highest levels of hazardousness. Those findings are similar to the

Table 7.2 Spanish municipalities with high or very high IH and the population exposed

Index of hazardousness PI	Municipality	Province	Population No. of inhabitants	% of inhabitants
5	Barcelona	Barcelona	1,503,884	29.0
5	Saragossa	Saragossa	614,905	11.8
4	Vitoria-Gasteiz	Álava	216,852	4.2
4	Prat de Llobregat (El)	Barcelona	61,818	1.2
4	Rubí	Barcelona	61,159	1.2
4	Sant Cugat del Vallés	Barcelona	60,265	1.2
4	Aranda de Duero	Burgos	29,942	0.6
4	Malaga	Malaga	524,414	10.1
4	Murcia	Murcia	370,745	7.1
4	Gozón	Asturias	11,074	0.2
4	Sevilla	Sevilla	684,633	13.2
4	Valencia	Valencia	738,441	14.2
4	Valladolid	Valladolid	316,580	6.1
Total population			5,194,712	100.0

7 The Case of Spain 123

ones produced by the methodology applied in the present work. Certain high risk areas, mainly due to the presence of the chemicals sector, are identified by both studies: Tarragona, Asturias, Cantabria (Torrelavega), País Vasco, Valencia, Galicia (Pontevedra), Castilla y León (Miranda de Ebro) and Aragón (Sabiñánigo/Monzón). However, some high-risk areas in the Greenpeace report were not identified as such by our methodology (Huelva, Campo de Gibraltar, Castilla-La Mancha (Puertollano), Madrid (Alcalá de Henares, Aranjuez and Getafe), and vice versa (Barcelona, Valladolid, La Coruña, Segovia, Baleares, Cordoba, Malaga and Granada). These results highlight the importance of applying specific methodologies to assess risk and its different dimensions using homogeneous variables that permit greater differentiation between zones.

4 Social Vulnerability in Spain

As for the other case studies, the methodology applied to characterise social vulnerability in Spain is as expounded in Chapter 3 of the present work. The selection of indicators is the result of a thorough comparison of the available sources in the different countries which were the subject of the study.

The source of the information used to assess social vulnerability is the National survey of Population 2001 (INE 2001) whose data can be accessed without charge via the website of the National Institute of Statistics (Instituto Nacional de Estadística). This website offers unrestricted access to databases of different levels of aggregation. As was the case for industrial hazardousness, the availability of information has allowed us to characterise Spain's social vulnerability on the municipal scale and disaggregate information was used for the census unit.

Table 7.3 depicts the data from the Spanish census corresponding to the indicators selected:

Table 7.3 Dimensions and indicators of social vulnerability in Spain

Dimension	Indicator	Definition
Demography	1. Transitory dependent population	Population under 14 years of age
	2. Definitive dependent population	Population over 65 years of age
	3. Single-parent households	Household of 1 adult responsible for at least 1 minor.
Economic capacity	4. Health cover	
	5. Literacy/education	Total number of illiterates
	6. Profession/employment	Profession of the person of reference – No. of unemployed per household
Standard of living	7. Dwelling	Average surface area of dwelling per inhabitant
	8. Access to drinking water	Number of households without running water
	9. Sewage treatment	Households without sewage treatment

The definitions of the indicators in the censuses of the different countries in the study do not always coincide, but they do show certain equivalence for each of the three dimensions of vulnerability that make up the index (demography, economic capacity and standard of living). The indicators used are presented in absolute values (number of inhabitants, number of households, surface area of the dwelling, etc.)

"Health cover" is a very important indicator of social vulnerability in the case of Latin America. However, in the case of Spain this indicator is not included in the national census since all residents enjoy free access to the public health service. The possibility of finding equivalent indicators of health was considered, but although data are available from the National Health Survey, they are samples that are representative on the scale of the nation as a whole or of the autonomous community, but not at greater levels of disaggregation. It was considered relevant to maintain the indicator, pointing out that the whole population of Spain has health cover, which indicates that they are better equipped to face situations of risk.

On the level of both indicators and dimensions, scales of 1 to 5 were established using natural breaks (1, "very low", 2 "low", 3 "medium", 4 "high", 5 "very high"), so that sub-indices of vulnerability have been calculated for each of these levels. In analytical terms, treating the information in this way allows us to identify the weight of each indicator or dimension in the calculation of the index of vulnerability for each unit of analysis (municipality or province).

As indicated in Chapter 3, the index of social vulnerability (ISV) is made up of the direct sum of the results of the sub-indices obtained for the 9 indicators selected. The results can be seen below in the municipal-scale map of Fig. 7.2 and in the Table 7.4, which identifies the municipalities with very high ISV, their resident population and the percentage that population represents in the whole province. The table also reflects the percentage of the nation's population residing in municipalities with very high ISV. 17% of Spain's population is shown to live in areas of very high ISV, which are spread over 7 of Spain's 8,108 municipalities in 6 different provinces. The mapping of ISV reveals that Madrid, Alicante and Malaga are the provinces with the largest surface areas that register very high ISV. In the cases of Madrid and Malaga, the resident population in these towns represents a high percentage of the provincial total (54 and 41%, respectively, whereas in Alicante this tendency is not so marked, with 33% of the population living in the area with a very high level of ISV).

In absolute terms, Madrid and Barcelona are the municipalities with very high ISV with the greatest resident population. Their high level of urban development explains the high concentrations of population compared to the rest of the country's municipalities. In the case of Barcelona, the combination of geographic and statistical information reveals that the area with very high ISV does not cover a wide area, but on the other hand it does account for 30% of the province's population.

In relative terms, the percentages of population living in municipalities with very high SV in the provinces of Malaga and Seville are 41% and 40%, respectively. The province of Madrid registers the highest percentage of population living in such areas (54%), all in the municipality of Madrid itself, the country's capital.

7 The Case of Spain

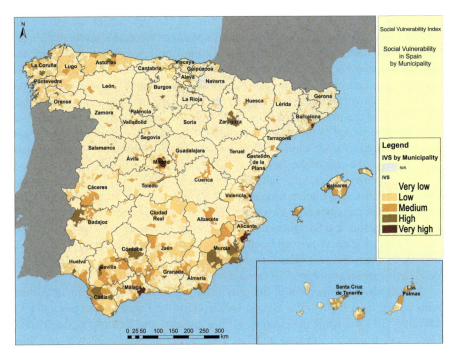

Fig. 7.2 Index of social vulnerability in Spain by municipalities
Note: the darker the more social vulnerability

Table 7.4 Spanish municipalities with very high ISV

Municipality	Province	Inhabitants	% of the province's population
Alicante	Alicante	28,4580	33
Elche		19,4767	
Barcelona	Barcelona	1,503,884	31
Madrid	Madrid	2,938,723	54
Malaga	Malaga	524,414	41
Seville	Seville	684,633	40
Valencia	Valencia	738,441	33
Total		6,869,442 (17% of the national population)	

In the provinces of Barcelona, Valencia and Alicante approximately one third of the population lives in municipalities with very high ISV.

Regarding the weighting of the factors, it can be seen that all the municipalities register very low ISV sub-indices for the indicators of health and dwelling. In Madrid and Barcelona no differences are observed in the weighting of the factors, with the sole exception of the access to drinking water indicator, whose weighting is high for Barcelona, whereas the other 7 indicators register very high levels.

Malaga and Seville register high levels in all the sub-indices except for the medium level of access to drinking water in Malaga. Valencia's levels of ISV are high or very high in 7 of the indicators, but the very low level registered for the indicator of transitory dependence is noteworthy. As far as the analysis of environmental risk is concerned, this indicator improves the situation of the population, but it also makes it vulnerable in other aspects related to its socio-economic sustainability. Regarding the two municipalities in Alicante, it seems clear that the high level of ISV is determined by the greater weighting of the indicators referring to access to services (drinking water and sewage), and to a high concentration of single-parent households in the case of the capital.

The 39 municipalities with high ISV represent less than 1% of all Spanish municipalities, but their population accounts for 18% of the national total. The mapping of ISV in municipalities and provinces reveals that the provinces of Murcia, Cádiz, Córdoba, Badajoz and Saragossa are those with the largest areas of high ISV. In the case of Murcia, three zones can be clearly distinguished with this high level of ISV, one on the coast and two inland. In proportion with the overall size of Santa Cruz de Tenerife, its areas of high ISV cover a considerable area.

As regards the proportion of the provincial population living in municipalities with a high level of ISV, the highest values are to be found in Saragossa (71%), Murcia (53%), La Rioja and Burgos (48%), Asturias and Santa Cruz de Tenerife (44%) and Baleares, Córdoba and Las Palmas (40%). Approximately one third of the inhabitants of the provinces of Almería, Cantabria, Castellón, La Coruña, Granada, Huelva, Ourense and Vizcaya live in municipalities with a high level of ISV.

In Saragossa, Murcia and Córdoba there is a high proportion of population affected and also larger areas with high ISV. On the other hand, Badajoz and Cádiz are among those municipalities where although large territorial areas have high ISV, there are low percentages of population living in such areas.

Bearing in mind the weighting of the factors of ISV, as in the case of municipalities with very high ISV, all those with high ISV register the lowest level for the sub-indices of health and dwelling. For the other sub-indices the values tend to be spread between 1 and 3 in most cases, and the sub-indices with the highest values are those of access to services (drinking water and sewage). Only in the case of Lorca is a high level observed for the sub-index of transitory dependence; in Gijón, Bilbao and Saragossa there is a high level of the definitive dependence indicator, and in Palma de Mallorca, Las Palmas de Gran Canaria and Saragossa the single-parent households indicator also has a high value.

Table 7.5 presents the data referring to the number of municipalities at each level of vulnerability and the relative weight of the resident population for each level expressed as a percentage of the country's total population. It also includes data that identify the relative weight of three aggregate levels of vulnerability (high, medium and low) expressed as percentages of the national totals.

The vast majority of municipalities have a very low level of ISV, and 97% of them, home to 41% of Spain's population, are within the low aggregate level of ISV. On the other hand, 23% of the population lives in areas with medium ISV which represent only 2% of the total number of municipalities. The map of ISV shows

7 The Case of Spain

Table 7.5 Municipalities and resident population according to levels of ISV

ISV	No. of municipalities	% of municipalities per level of ISV	% of national population	% of population per level of ISV
1	7,179	97	24	41
2	692		17	
3	191	2	23	23
4	40	1	18	36
5	7		18	
Total	8,109	100	100	100

that the largest areas with this level of vulnerability are to be found in the provinces of Murcia, Albacete, Cáceres, Jaén, Sevilla Cádiz, Huelva, Granada and Almería. Finally, a small number of municipalities, a mere 1% of the total, present high levels of ISV, but these areas account for 36% of the total population. The urban make-up of the municipalities in this last category can be deduced from the high proportion of them (25 out of 47) that are provincial capitals.

The areas with very low ISV reflect very low concentrations of population, and it is noteworthy that only 24% of the population live in such a large number of municipalities (7,719); indeed it can be seen from the map that this category covers the largest areas of territory.

5 Combined Risk Due to Hazardousness and Social Vulnerability in Spain

This section presents the results derived from the analysis of indices of combined risk (IR) for the whole of Spain at municipal level. As explained in the chapter on the methodology and theoretical framework , the result of combining the indices of industrial hazardousness (IP) and social vulnerability (ISV) has allowed us to identify the so-called "hot spots", i.e. areas with high (IR = 4) or very high (IR = 5) combined risk. Figure 7.3 presents the map of combined risk for the whole of Spain. Following a traffic-lights colour code, "hot-spots" are represented in red (IR = 5); and orange (IR = 4); areas with medium risk (IR = 3) in yellow; areas with low risk (IR = 2) in light green and finally areas with very low risk are shown in darker green (IR = 1).

By combined analysis of the geographic information in Fig. 7.3 and the aggregate statistical information in Table 7.6, it can be seen that 98% of Spanish municipalities present the lowest levels of combined risk (IR = 1, IR = 2) and that the majority of these fall into the very low IR category. These areas are inhabited by 50% of the country's population, while 10% live in the 1% of municipalities that register a medium level of IR.

The remaining 1% of municipalities presents high or very high IR, but it is noteworthy that 40% of the Spanish population live in such areas. The largest areas affected by this level of IR are to be found in Murcia, Córdoba, Alicante, Cádiz, Madrid and Saragossa. Other hot spots which cover less area can be found

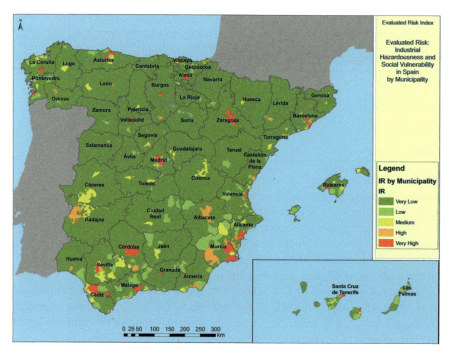

Fig. 7.3 Map of evaluated risk in Spain by municipalities

Table 7.6 Municipalities and resident population according to levels of IR

IR	Municipalities	% of municipalities per level of IR	% of national population	% of population per level of IR
1	7,764	98	36	50
2	193		14	
3	79	1	10	10
4	35	1	7	40
5	37		33	
Total	8,108	100	100	100

in provinces such as Barcelona, Malaga, Sevilla, Granada, Valladolid, Burgos, La Rioja, Valencia, Castellón, Álava, Vizcaya, Asturias, La Coruña, Pontevedra, Cantabria, Baleares and Santa Cruz de Tenerife.

Table 7.6 reflects the levels of combined risk (IR) for Spanish municipalities together with the resident populations and the percentages that they represent of the provincial and national totals.

A total of 37 Spanish municipalities form 24 different provinces present a very high level of IR, and they account for 33% of the country's population. Madrid is the only municipality with very high IR in the province of the same name, and this city is home to 7% of the total Spanish population. In second place regarding population size comes the province of Barcelona with 6% of the national population, most of which resides in the capital. Valencia and Seville each have two municipalities that register very high IR and the sum total of each of their populations represents approximately 2% of the national total. None of the remaining provinces with very high IR has a population that reaches 2% of the total.

Regarding the provincial proportions, the population living in areas with very high risk (IR = 5) represents 43% of the total population of the 24 provinces in question. It can be seen that Álava, Valladolid and Saragossa have the highest percentages of provincial population living in such areas, with 76, 64 and 71%, respectively. A second group, consisting of Barcelona, Burgos, Madrid, Murcia, La Rioja and Seville, have roughly half of their population living in such areas. This percentage drops to around 40% for Asturias, Baleares, Córdoba and Las Palmas, while Cádiz and Santa Cruz de Tenerife are the provinces with the lowest percentages of population living in areas with very high combined risk.

6 Case Studies in Madrid and Seville Using the Census Block Group or Census Unit as Unit of Analysis

One of the aims of this work was to analyse levels of risk, taking into account social vulnerability and industrial hazardousness in greater depth than at municipal level. To this end, two case studies were selected from among the municipalities with high risk. it was decided to choose two provincial capitals whose indices of risk due to vulnerability (ISV) and hazardousness (IP) present differing values, namely Madrid (IP = 2 IVS = 5) and Seville (IP = 3 IVS = 5).

By applying the methodology on the scale of the census unit, we have been able to draw up the following maps of social vulnerability, environmental risk and combined potential risk in order to identify hot spots on this detailed level in the two cases studied. Table 7.7 provides an overall description of the data corresponding to Madrid and Seville.

The maps of vulnerability (Fig. 7.4), hazardousness (Fig. 7.5) and risk (Fig. 7.6) for both municipalities are presented below.

Table 7.7 Descriptive data of the municipalities analysed on the scale of census units

Madrid		Seville[2]	
Total inhabitants	2,907,036	Total inhabitants	485,947
Total households	1,072,780	Total households	166,421
Total firms analysed	9,232	Total firms analysed	6,300
Total districts	21	Total districts	6
Total sections	2,326	Total sections	383

---

[2] The cited data for Seville only take into account urban census units. The total population is 684,633 and the total number of households is 226,621.

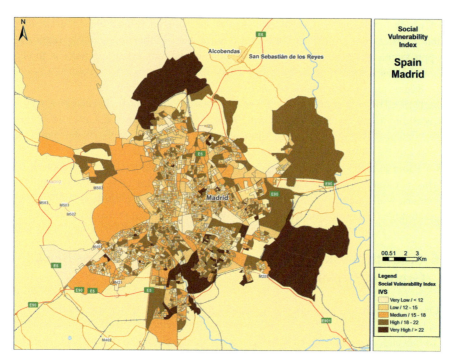

Fig. 7.4a Map of social vulnerability in Madrid by census unit

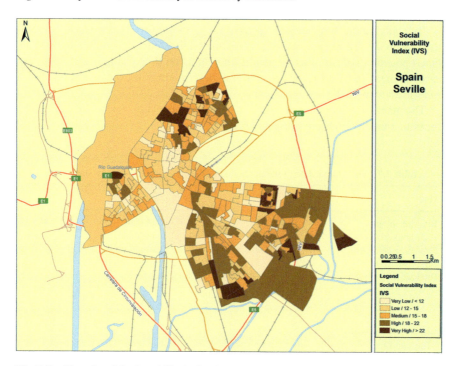

Fig. 7.4b Map of social vulnerability in Seville by census unit

7 The Case of Spain

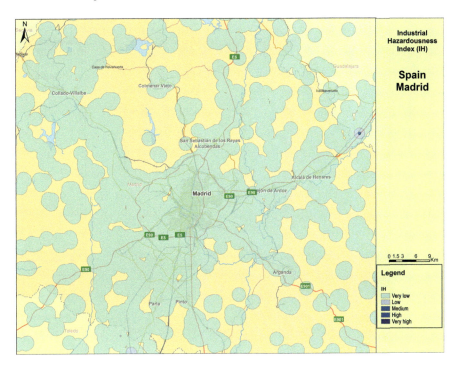

Fig. 7.5a Map of industrial hazardousness in Madrid

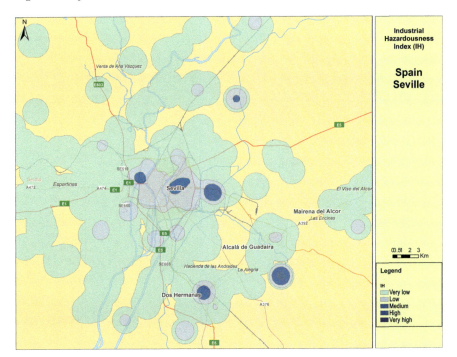

Fig. 7.5b Map of industrial hazardousness in Seville

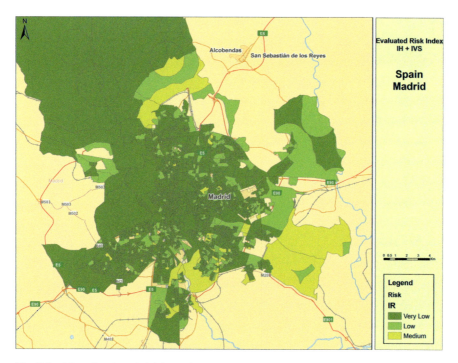

Fig. 7.6a Map of evaluated risk in Madrid by census unit

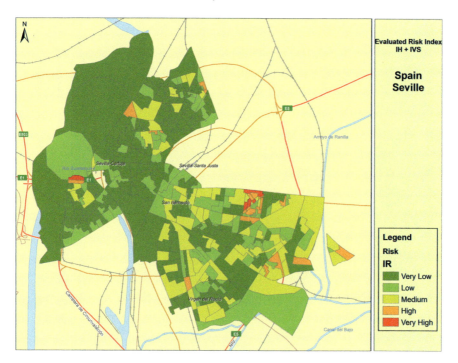

Fig. 7.6b Maps of evaluated risk (IR) in Seville by census unit

7 Social Vulnerability in Madrid and Seville

In Figs. 7.4a and 7.4b the darker the shade of brown the higher the IVS, In the case of Madrid, the map of social vulnerability shows five large vulnerable areas located on the outskirts of the city, three with very high ISV and two with high ISV. A further three smaller areas with a high level of SV can be distinguished also on the outskirts. The data reveal that of Madrid's 2,326 sections only 49 (spread over 14 districts) register very high SV, while 289 (spread over 21 districts) register a high level. The total number of sections included in these two categories (338) accounts for 14% of the total, whereas the population and the number of households both represent 20% of the totals

The map of Seville reveals three medium-sized areas with very high ISV located to the north, east and south as well as several smaller ones dotted around the outskirts of the city. The large size of the area of high level of ISV to the east of the city is worthy of note, and particularly the area of very low vulnerability that can be seen around the Guadalquivir River. Of the 330 sections considered, 27, spread over 5 districts, register very high ISV and 57, spread over 6 districts register a high level. The census reveals that 21% of the inhabitants and households of the municipality are to be found in these areas.

8 Industrial Hazardousness in Madrid and Seville

According to the categories established earlier for the whole of Spain – where darker shades of blue correspond to higher IP – the map of industrial hazardousness for Madrid (Fig. 7.5a) shows that the levels of Industrial hazardousness are at the lower end of the scale. This is consistent with the comments expressed above to the effect that Madrid is not notably affected by environmental risk due to industrial hazardousness.

As far as Seville is concerned (Fig. 7.5b), analysis of the data leads to the identification of several census units that present higher levels of contamination; though in no cases does this level surpass the medium category as defined in the methodology. 28 of the 330 census units that make up Seville register levels of medium hazardousness. Some 35,932 inhabitants live in these areas in a total of 12,322 households.

9 Evaluated Risk in Madrid and Seville

Figure 7.6 presents the maps of combined risk for Madrid and Seville. It would seem significant that none of the areas of Madrid registers high or very high levels of evaluated (potential) risk (IR). The highest value registered is medium, and this in only 42 of the 2,326 census units into which the municipality is divided (spread over 14 of the 21 districts). The population living in these areas represents 3% of the municipal total and 2% of the provincial one, and the same holds true for the percentage of households. In short, the analysis reveals that 97% of the resident population and 97% of the households in Madrid register low or very low levels of IR.

In Seville on the other hand, the results show that 4 of the 330 sections register a very high level of IR, while 13 register a high level. These 17 "hot-spots" contain 4% of the total population and households. To be precise, sections 3, 5 and 7 (district 4) and section 17 (district 6) are those that register this very high level of potential risk. In total, 26,684 inhabitants of Seville live under high or very high environmental risk

10 Concluding Remarks

This chapter aimed to apply a methodology that would allow the identification of areas of Spain of evaluated high risk due to the combination of hazardousness and social vulnerability. As well as identifying municipalities on a national scale, the methodology has permitted two case studies on the scale of census units to be carried out in Madrid and Seville.

At local level, the analysis has made it possible to notice that social vulnerability levels are higher in the surroundings of provincial capitals and that, although not a great number of municipalities with high social vulnerability levels have been presented, the highest risk cases are in the main capital cities (Madrid, Barcelona, Valencia, Seville and Malaga). Regarding industrial hazardousness, although this study has used a working methodology aimed at evaluating, not only the impact of certain industrial sectors, but the sum of all the activities, the results show how the areas with higher risk due to hazardousness, are those which have a more important industrial concentration, more than with a general aspect of concentration of industries. Joint analysis of both social vulnerability and industrial hazardousness, reveals a large number of municipalities under high risk. To put it briefly, social vulnerability can be connected to the more densely populated areas (especially provincial capitals), while industry hazardousness is more clearly marked by the location and concentration of specific industries rather than by a cumulative effect of industrial concentration

Within the sphere of the census unit, these conclusions are not applicable, since the detailed analysis is directed towards the identification of specific areas at risk.

The results suggest that no hot spots are located in Madrid, whereas four census units in Seville have been revealed as hot spots with a very high potential risk. Although there are areas of high SV in both cities, they are larger in Madrid. The difference can be explained by the lower degree of industrial hazardousness in this city. There may be several reasons for this: the historical characteristics of the Spanish industrial sector, the high value of land in Madrid and competition for its use on the part of services firms, specific legislation regulating distances at which high-risk firms must be located (Real Decreto 1254/1999 or Ley de Accidentes Mayores). All these factors may explain why the most contaminant industries, which require greater space and a more stringent control of the inherent risks of serious accidents, have sought alternatives to the country's capital for their location. This means that we can speak of decentralisation of the Spanish industrial sector.

The results show that certain hot spots exist. These are areas of particular concern due to the dual potential risk of vulnerability and hazardousness, and they merit analysis in greater depth, both at municipal level and at the level of census units, combined with a more qualitative approach.

Our methodology is particularly useful to aggregate the impacts of different activities that affect a given territory. The tools for the analysis of environmental risk proposed by the Technical Commission for the Prevention and Reparation of Environmental Damage (Models of Environmental Risk Reports, Tables of Criteria and Methodological Guide) refer to the intrinsic risk of a given activity. Nevertheless, we believe that a further step needs to be taken, aggregating the individual risks of the different entities in a similar approach to the one here proposed.

When the application of Law 26/2007 of Environmental Responsibility and regulation (RD 2090/2008) have generated enough additional information of the industrial hazardousness of each sector, it may be possible to refine our model further and to use as initial data more precise parameters to quantify potential risk. Our model only takes into account quantitative aspects of industrial hazardousness, but it does not go into the exact nature of each risk and their combined effect on a given area, e.g. the risk of fire in a certain area combined with the risk of generation of toxic gases.

References

Greenpeace. (2008). *Informe: Contaminación en España 2008*. Barcelona: Greenpeace España.
INE-Instituto Nacional de Estadística. (2001). Censo Nacional de Población, Madrid.
INE-Instituto Nacional de Estadística. (2005). Anuario Estadístico de España, Madrid.
INE-Instituto Nacional de Estadística. (2008). Encuesta Nacional de Inmigrantes, Madrid.
INE-Instituto Nacional de Estadística. (2009). Anuario Estadístico de España, Madrid.
Ministerio de Medio Ambiente. (2008). Perfil ambiental de España 2008.

Chapter 8
Concluding Remarks

Jerónimo de Burgos-Jiménez, Diego A. Vazquez-Brust, José A. Plaza-Úbeda, and Claudia E. Natenzon

Abstract This chapter provides a summary of the conclusions drawn from all previous chapters, discusses its implications and provides policy recommendations. Limitations are stated as well as directions for further research suggested. The chapter emphasises that on the whole, the results obtained have confirmed the usefulness of our conceptual and methodological tools to assess the risk due to environmental deterioration to which the population of a given territory is susceptible. In the three country-case studies – Spain, Argentina and Bolivia – the comparison of spatial patterns and indicators in two different scales of analysis (regional and 'census unit') has identified significant geographic differences in terms of the distribution of vulnerability and hazardousness. In other words, irrespective of their degree of economic development, the three countries present scenarios of high vulnerability and/or hazardousness. This analysis constitutes the first step towards the management of risk and may help in the design of preventive measures. However, solving these problems implies that decision making entities must be capable of acting on the causes of risk.

Keywords Environmental risks · Industrial hazardousness · Environmental justice · Spatial analysis · Hot-Spots

This book presented the results of a three year international collaborative project lead by Almeria University, University of Buenos Aires and Cardiff University. 'Hot-Spots': mapping social vulnerability and environmental risk. The project assessed social vulnerability and environmental threats posed by economic activities.

The book has expounded and validated an innovative methodology to assess the risk to which a given territory is susceptible by adopting a multidisciplinary approach. It did so by combining environmental, socio-economic and geographical concepts to construct new spatial and technical indicators that assess the Social Vulnerability and Industrial Hazardousness of a given territory. Mapping the indicators in a geographic information system facilitated the assessment of potential

---

J. de Burgos-Jiménez (✉)
Department of Business Administration, University of Almeria,
Almeria 04120, Spain
e-mail: jburgos@ual.es

environmental deterioration caused by industrial activities located in certain areas. Two scales of analysis were used to make the assessment: a regional one that identifies average risk over large administrative areas such as provinces, and a more detailed scale, the *census unit, to* determine the local distribution of risk and contaminant industries. A census unit is the smallest area for which census data is collected in a country, typically containing between 0 and 1,000 people and up to 250 housing units. The methodology applied identifies geographical variations of risk levels within the same country and is also useful for comparing data across different countries.

As explained in Chapter 2 on the conceptual framework, evaluated Environmental Risk is basically a reflection of the information provided by spatial and technical indicators. In order to obtain a more complete vision of Environmental Risk, these aspects must be complemented by policies, actions and values which refer to the dimensions of governability and uncertainty. Furthermore, the values of evaluated risk reflect the potential hazard of environmental damage that may be caused by economic activity on the territory analysed and the population living there.

The great differences in levels of social vulnerability and poverty in Latin America that are highlighted in Chapter 3 (e.g. housing conditions, access to running water, sewage, education, health coverage, single parent families, informal employment) underline the importance of considering these aspects alongside environmental hazardousness in order to determine the evaluated risk according to the methodology, which is described in detail in Chapter 4.

This methodological approach to calculating evaluated risk has been applied to three countries which share the same language, but whose socio-economic situations vary greatly, namely Spain, Argentina and Bolivia. The scenario in Spain differs greatly from that in the Latin American countries regarding policies, actions and values, and the results obtained must therefore be interpreted differently. The strong institutional pressure that Spain has been under since it joined the European Union has led to an increase in social concern for safety at work and protection of the environment. In particular, the high levels of environmental demands established by the EU norms to which Spain is subject mean that the gap between evaluated risk and real or managed risk in this country is greater than in the other two: Argentina and Bolivia. As pointed out in the analysis of governability in Latin America (see Chapter 4), environmental legislation has had little impact on improving the performance of Latin American firms, while the same cannot be said for Spain. It may be, therefore, that the real risk in Spain is considerably less than has been evaluated in this work, since there is already a high degree of commitment to using State resources to supervise the enforcement of specific measures of environmental protection and of risk control imposed by the regulations (e.g. set of norms on environmental protection or the risk of major accidents). Nonetheless, we should also consider that environmental values have been taken on board less in Spain than in many other European countries, and some Spanish organisations find it difficult to comply with current environmental norms.

8 Concluding Remarks

Apart from implying differences in interpreting the results of the different countries analysed,[1] this heterogeneous socio-economic context has had a considerable bearing on the application of the methodology. On the one hand, access to the necessary data to construct the indicators for the assessment of exposure, vulnerability and environmental hazardousness has differed greatly from one country to the next. On the whole, as the data were not the same in all the countries, similar indicators were sought to reflect the same dimension based on the available data. For instance, the indicators of economic capacity and standard of living tend to differ regarding the indicators of literacy/education, work/occupation and dwelling. In all cases equivalent indicators were agreed on in the multilateral meetings.

On the other hand, the spatial/political division also differs in the three countries analysed. Spain is made up of Autonomous Communities, provinces, municipalities and localities, whereas Argentina is divided into provinces, municipalities and localities and Bolivia into departments and capital cities of departments. It was therefore finally agreed that department and municipality should be regarded as the same, and that the smallest unit of analysis would be the census unit. There have also been differences in the quality and quantity of the data compiled in the different countries. Generally speaking, it could be said that these aspects depend on the level of economic development of the country and the existence of specialised institutions to compile and publish these data (for example Spain's INE, or National Institute of Statistics). Data collection has been easier in Spain, where we have been able to adapt our indicators to data from secondary sources (principally the INE and the SABI database), but it has proved very problematic in the case of Bolivia. In this country, the process of information gathering has been a complex, laborious task. As it was impossible to gain access to street maps of the municipalities of Santa Cruz de la Sierra and Sucre, the industries could not be placed on the map automatically but rather the researchers had to carry out this task manually. In addition, some proxy measures had to be taken and adapted to the existing data; also it proved impossible to reach the level of census unit as there is no data at this level of disaggregation.

Applying this methodology to both developing and developed countries may condition the treatment of some indicators or dimensions, as may the geographical characteristics of the countries. For instance, in the case of Spain we assume that the whole territory is within half an hour's travelling distance from a medical centre or hospital, and so health cover is universal. However, this is a very important factor when considering social vulnerability in Latin America, and so it must be quantified as a dimension of vulnerability in Bolivia and Argentina.

On the other hand, the choice of the three countries analysed and the results obtained allow different suggestions to be made regarding the remaining Ibero-American countries. From the methodological point of view, the first implication

---

[1] As is indicated in the introduction, the difficulty in accessing information, in terms of both cost and time, have meant that it has been impossible to carry out the initial project that had foreseen the study of risk in other countries, such as Venezuela and Brazil.

of this study refers to the differences in the availability of information in the different countries chosen. It has proved more difficult to gain access to information in those countries with a lower level of development (in particular in Bolivia), where public institutions lack the necessary resources to have the same level of available information as more developed countries (e.g. Spain). This should be borne in mind if further studies are carried out in the context of countries in similar circumstances. Though political issues, democratic tradition and the stability of the institutions also have a bearing on this issue, the difficulty in obtaining information and the depth of that information can be expected to be of a similar nature in countries with similar levels of development.

The varying socioeconomic situation of the three cases analysed (Spain, Argentina and Bolivia) also allow to use the cases as examples and referents to be applied in other countries with similar socio-economic configurations. In other words, the results of the present work can be used as templates to develop initiatives in countries in similar socioeconomic circumstances to the three cases studied. More concretely, countries similar to either Spain, Argentina or Bolivia in terms of relationship between the level of economic development, the risk due to social vulnerability and environmental deterioration and the structural conditions that reinforce situations of the environmental injustice.

For instance, the results in Argentina, a country with a high level of development and low poverty in the Ibero-American context, would seem to suggest that the characteristic problems and situations of Argentina regarding social vulnerability and the environment would be similar in other countries with a similar level of economic development, degree of industrialisation, distributive inequality or legislation (e.g. Brazil, Chile, Uruguay, Costa Rica, Mexico or Portugal). Likewise, for the case of Bolivia the combined risk due to social vulnerability and industrial hazardousness may well be similar in countries such as Honduras, Ecuador, Nicaragua, Guatemala, Peru or Colombia. The case of Bolivia highlighted, that countries with relatively low levels of industrial activity but high levels of inequality can still have severe localised situations of high risk, which are only evident when the analysis is carried on at the level of census unit.

Finally, the case of Spain confirmed a substantial structural gap in terms of equality, vulnerability and environmental justice between the country and the rest of Iberoamerica. Despite the history, culture, economic relations Spain is structurally more akin to southern Europe than it is to Latin-America. This gap can be seen at the macro level. Mexico, Brazil or Argentina are not that far from Spain in terms of GDP but they clearly lag behind Spain's in terms of equality and social cohesion (Spain's GINI coefficient is half than Uruguay's, the latter being the most equalitarian Latin-American countries). There are also striking differences at the level of the Census Unit. While in Argentina and Bolivia vulnerable populations and polluting firms tend to converge in the same areas. The analysis in Madrid and Seville unveil the existence of areas with high social vulnerability, but none or very few of them are exposed to industrial hazards. Allegedly, the results from Madrid should not be a surprise, since Madrid is an administrative city with few industry and many

8 Concluding Remarks

services. However, although the same can be said of Sucre – administrative capital of Bolivia –, this city has large numbers of people exposed to high risk. The little industry that Sucre has is located in areas inhabited by highly vulnerable populations. Thus, the city is polarised between marginalised communities living in Hot-Spots and large residential areas with very low risk.

Despite all these considerations, it can be concluded that the social variety (chiefly geographic, economic and distributive) of these countries requires certain flexibility regarding the application of this methodology. Moreover, the construction of these indicators can be perfected by incorporating additional information or by improving the tools of aggregation or weighting. For instance, although in Argentina specific indicators were available for weighting the environmental complexity of industries that are prone to major accidents, the same cannot be said for Bolivia, and limited resources meant that they could not be obtained for Spain. Consequently, in order to maintain the criterion of comparability, these indicators were applied in Argentina only in the case of the study at the level of census unit, whilst the analysis on the level of departments in the three countries considered the same formulae used to calculate indicators based on estimation of emissions. With greater available resources the analyses of Spain, and indeed of EU countries subject to the same norms, could incorporate specific weighting qualifying the industrial sites prone to major accidents in their different categories. These progressive adjustments to the model would help to identify critical points in each territory with greater precision.

The methodology defines vulnerability as a multi-dimensional construct whose evaluation is based on a set of indicators that measure aspects of the social reality that expose different situations of weakness or fragility of the social groups studied which make them better or worse prepared to face up to the negative impacts arising from hazardous processes associated with business activity.

The methodology used is more powerful than those tools that assimilate vulnerability with situations of poverty, as it allows us to identify situations of susceptibility to hazards that go beyond the level of income (for instance age or access to infrastructure).

Mapping the data to assess situations of hazardousness arising from cumulative negative impacts of firms was an innovative approach. It allowed the identification of hazardousness due, not only to large industries, but also to geographical clusters of numerous small industries whose individual hazardousness was insignificant individually (and therefore less regulated or controlled), but whose combined emissions may have constituted a greater threat than that of a single large firm. This is particularly important because small firms or those posing little hazard are much less visible and face less scrutiny than large firms, and so they are not only less regulated and controlled (less governability), but also less is known about their cumulative effect on the population (greater uncertainty). Consequently, there are usually fewer policies, actions and values directed at managing risk generated through geographical proximity to small emitters compared to large industries. The application of our methodology provides diagnostic tools to overcome this problem in distributing institutional resources for managing risk.

One of the added values of this work stems from the fact that it is carried out homogeneously for the whole territory and it provides a single aggregate value of the evaluated risk. This knowledge constitutes the first step towards control, but the active involvement of social agents, and especially of government, (in their respective scope of competence) is essential if we are to progress towards the prevention and reduction of risk. Without such measures the analysis of the data is of very limited value in progressing towards sustainable development. Our methodological approach is valid to identify and quantify potential risks: some high levels of risk may anticipate catastrophes, and the analysis of risk may help us to design preventive measures. However, solving these problems implies that those entities with the power to take decisions must be capable of acting on the causes of risk.

On the whole, the results obtained have confirmed the usefulness of this methodology to assess risk due to environmental deterioration to which the population of a given territory is susceptible. In the three countries studied the indicators have identified significant geographic differences in terms of the distribution of vulnerability and hazardousness. In other words, irrespective of their degree of economic development, the three countries present scenarios of high vulnerability and/or hazardousness. These 'Hot Spots' have mainly been identified in large cities where there are higher concentrations of firms and of people (usually associated with rapid growth that compromises the appropriate adaptation of the region). Nevertheless, the level of development and the predominance of the tertiary sector in the economy may condition the level of risk. This would appear to be illustrated in the analysis on the scale of census unit of the municipality of Madrid: although there are areas with a high level of social vulnerability, the presence of very few manufacturing firms and many services ones (of less environmental hazardousness) means that the level of risk is acceptable.

The effort involved in compiling, sorting and treating the data becomes more intensive the more specific the level of analysis. It is therefore difficult to carry out an analysis at the level of census unit for the whole territory due to questions of both time and resources. In this sense, the methodology can be considered hierarchical, so that at the most aggregate level the analysis sheds light on where resources need to be focussed: namely in those territories in which a high level of risk was evaluated at the preceding levels of aggregation.

This work provides a single aggregate value of overall risk. This analysis constitutes the first step towards the management of risk and may help in the design of preventive measures. However, solving these problems implies that decision making entities must be capable of acting on the causes of risk.

This approach was implemented at country and municipal levels to obtain results for both the municipal and census units. These different levels of analysis will facilitate the pinpointing of efforts in planning and controlling evaluated environmental risk at different levels of administrative decisions: country or region, municipality and locality or even specific entities (such us firms or groups of firms).

National Government should pay greater attention to aggregate data to control risk at different administrative levels. It could even be used to decide on the allocation of resources for vigilance and for plans to prevent or control risks to the population (e.g. incentives to decentralise large agglomerations, etc.).

8 Concluding Remarks

The analysis of the more detailed data (on the scale of the census unit) may prove especially useful for responsible for spatial planning: town planning schemes, or the concession of permits for industrial activity and for building. The detailed data may also be useful in planning the safety and protection of the population (e.g. security forces, Civil Protection, fire service, hospitals, etc.). As such, these Hot Spot maps could be useful for designing and establishing emergency protocols or for planning simulation exercises. For instance, in Spain, Civil Protection have data identifying the establishments that constitute potential environmental hazards (stipulating for each one the type of risk involved), and they coordinate the action to be taken in each case, but less attention is afforded to the cumulative hazardousness of individual low-hazard firms or to the risk due to the exposure of highly vulnerable populations. The spatial approach of this work allows us to consider these effects on the environment and their effect on the population as a whole.

The publication of the analysis of these data and of the disaggregate maps may also prove beneficial to firms. On the one hand, they make them aware of the impact that their activity generates on the territory and they allow them to foresee possible consequences. On the other, they may influence firms' decisions on where they should establish their activity: firms should prefer to be located away from high risk zones in order to avoid greater scrutiny from the public and greater costs of environmental responsibility.

The application of the methodology here presented at these levels of analysis should not be considered a closed issue. It could be extended to a larger geographical area, providing the body responsible for its control has the power to allocate resources to adopt corrective measures. This could be done for instance in the European Union countries in Europe and in America to the MERCOSUR countries.[2] The European option is particularly interesting, as it could be used as a complementary criterion for allocating the structural funds[3] that the EU distributes among its least prosperous regions.

On the other hand, the application of this methodology, based on analysis with secondary data, only allows us to reach the census unit as the smallest unit of analysis. However, other complementary approaches, such as the in-depth study of firms in areas of high evaluated risk and interviews with the main stakeholders may provide new perspectives of risk and its management.

---

[2] MERCOSUR is a full customs union founded in 1991 between Argentina, Brazil, Paraguay and Uruguay. Bolivia, Chile, Colombia, Ecuador, and Peru have associated member status.

[3] These structural funds are: ERDF, ESF, EAGGF and FIFG; they are currently directed at improving the economic and social cohesion among member countries, but they could also be extended to cover aspects of environmental risk. The first two funds offer the best possibilities, while the latter two are specifically intended for agricultural and fishing activities. The European Regional Development Fund (ERDF) basically contributes to helping the least developed regions and those which are undergoing processes of economic reconversion or which suffer structural difficulties and the European Social Fund (ESF) intervenes mainly in the context of European employment strategy.

In this sense, the project also prompted the mapping of community-business partnerships that reduce vulnerability and environmental deterioration in the identified Hot Spots. This revealed the importance of citizenship and personal engagement as well as companies' proactivity to open institutional spaces to generate bottom-up projects. For example, an initiative to break poverty traps stimulates creative thinking in children from some of the most critical hot-spots areas mapped in Chapter 6. It organises workshops where vulnerable children create 'ideal worlds': drawing characters, recording sounds and writing scripts, which are then captured in three-dimensional projection loops. It started as a voluntary project led by a local communication expert. A pilot was funded by the telecommunications giant *Telefonica* through its *Telefonica Foundation* and it is currently maintained by Buenos Aires Municipality as a tool to engage vulnerable children and enhance their wellbeing; with the pioneering children acting as guests in a new series of workshops.

Index

A
Adaptation, 21–22, 119, 146
Adaptive capacity, 21–22
　definition, 21
Argentina, 7, 12, 23–24, 43–74, 91–115, 138–141
　Almirante Brown, 94, 106, 112
　biodiversity, 6, 93
　Capital Federal, 101
　　See also Ciudad Autónoma de Buenos Aires (CABA)
　Catamarca, 101, 103–104, 108
　Chaco, 101, 103, 106, 108
　Chubut, 94, 101, 103, 106, 108
　Ciudad Autónoma de Buenos Aires (CABA), 94, 106
　　See also Capital Federal
　Cordoba, 106, 108
　Corrientes, 101, 103, 106
　crises, 92
　economy, 92
　El Dorado, 104
　Entre Rios, 101, 103–104, 106
　environmental activism, 93
　environmental indicators, 23, 29
　environmental problems, 93
　environmental regulation, 93–94
　Federación, 104
　Florencio Varela, 94, 106
　Formosa, 101, 108
　General Pueyrredón, 104, 106
　General Roca, 104, 106
　General San Martín, 94, 104, 110, 112, 114
　geography, 91
　Guaymallen, 104
　industrial hazardousness, 94–97, 101–104, 106–113, 115
　Jujuy, 101, 103
　La Capital, 106
　La Matanza, 94, 104, 106, 109, 112–114
　Lanus, 94, 104
　La Plata, 104, 106, 108, 111
　La Rioja, 101, 103–104, 108
　Lomas de Zamora, 94, 106, 112
　Mendoza, 101, 103–104
　Merlo, 94, 106, 112
　Misiones, 101, 103–104, 106, 108
　Moreno, 94, 106, 112
　Neuquen, 101, 103, 106, 108
　resources, 92–93
　Rio Negro, 103–104, 106, 108
　Risk, 92–94, 103
　Rosario, 104, 106, 108
　Salta, 101, 103, 106
　San Juan, 101, 103–104
　San Luis, 101, 103, 108
　Santa Cruz, 101, 103, 108
　Santa Fe, 101, 103–104, 106, 108
　Santiago, 101, 103, 106
　social vulnerability, 92, 94, 96, 99–100, 104–109, 111–114
　social vulnerability indicators, 36–39, 42–43, 50, 52
　Tierra del Fuego, 101, 103, 108
　Tres de Febrero, 94, 104, 109, 114
　Vicente López, 94, 104, 109–110, 114
Asunción Declaration, 11

B
Bolivia, 7, 11, 24–25, 44–49, 69–90, 138–141, 143
　biodiversity, 70–71
　Cobija, 70, 76–77, 79, 81
　Cochabamba, 70–71, 75–77, 79, 89
　ecological problems, 71
　environmental indicators, 23, 29
　environmental regulation, 72
　geography, 91–92

Bolivia (*cont.*)
  industrial hazardousness, 74–84, 87–89
  industry, 93
  La Paz, 70–71, 73, 76–77, 79, 89
  Oruro, 70, 76–77, 79
  Potosí, 70, 76–77, 79
  poverty, 71
  resources, 70, 72
  risk, 74–76, 79–81, 87–89
  Santa Cruz de la Sierra, 70, 73–77, 79, 81–89
  social vulnerability, 74–76, 78–81, 84–89
  social vulnerability indicators, 36–39, 42–43, 50, 52
  Sucre, 70, 73–77, 79, 81–89
  Tarija, 70, 76–77, 79
  Trinidad, 70, 76–77, 79
Brazil, 24–25, 44–49, 92, 94, 139–140, 143
  environmental indicators, 23, 29
  social vulnerability indicators, 36–39, 42–43, 50, 52

C
Caribbean countries, 40–43
  social vulnerability indicators, 36–39, 42–43, 50, 52
Census block, 50–52, 94, 129–132, 138–143
  See also Census unit
Capital Federal, 101
Census unit, 5, 13, 89, 112, 114, 123, 129–132, 137–143
Chile, 24–25, 44–49, 94
  environmental indicators, 23, 29
  social vulnerability indicators, 36–39, 42–43, 50, 52
Chimney, 95–96
Ciudad Autónoma de Buenos Aires (CABA), 94, 106
CO2 emissions, 23, 25
Cobija, 70, 76–77, 79, 81
Cochabamba, 70–71, 75–77, 79
Colombia, 24, 44–49
  social vulnerability indicators, 36–39, 42–43, 50, 52
ConoSur, 23
Corporate Social Responsibility (CSR), 27
Costa Rica, 24–25, 44–49
  environmental indicators, 36
  social vulnerability indicators, 36–39, 42–43, 50, 52

D
Dasgupta and Wheeler, 65, 92, 95–97
Development, 1–3, 5–6, 9–12, 93, 98, 118–119, 124, 139–140, 142–143

E
Economic Growth-Environmental deterioration, 25, 29, 138, 140, 142, 144
Education, 36, 38–41, 44–45, 51
El Salvador, 24, 44–49
  environmental indicators, 23, 29
  social vulnerability indicators, 36–39, 42–43, 50, 52
Emissions, 20, 23, 25–26, 29, 61, 64–65, 95–97, 101, 103
  factor, 61, 96–97
    See also GINI coefficient
  per industrial sector:
    apparel, 97
    basic foodstuffs, 97
    beverages, 97
    ceramics, 97
    chemicals industry, 97
    computing and machinery, 97
    electrical appliances, 97
    footwear, 97
    furniture, 97
    glass, 97
    iron and steel, 97
    leather products, 97
    metal products, 97
    non-ferrous, 97
    oil products, 97
    oil Refining, 97
    other chemicals, 97
    other foodstuffs, 97
    other manufacturers, 97
    other non-metallics, 97
    paper, 97
    printing, 97
    professional equipment, 97
    rubber and plastics, 97
    textiles, 97
    tobacco products, 97
    transport equipment, 97
    wood products, 97
    See also Polluting particles
Empirical procedures, 22–23, 108–109
  See also Methodologies
Environmental awareness, 4, 26
Environmental damage, 3–4, 18–21, 25–26, 55–56, 138

Environmental deterioration, 2–4, 24–25, 28, 138, 140, 142, 144
Environmental hazard, 5, 6, 11, 13, 18, 29, 38–39, 109, 138–139, 142–143
Environmental justice/injustice, 13, 109, 112–113, 140
Environmental management, 25, 93–94
Environmental quality, 3, 11, 119
Environmental regulation, 3, 16, 20, 55–56, 119
   Argentina, 54, 56, 61, 64
   Bolivia, 54, 69–90
   Spain, 117–135
Environmental Risk, 10, 12, 15–30, 53–57, 126, 129, 133–135, 137–138, 142–143
   urban areas, 56, 92, 109
Environmental stress, 20, 26
Equator
   environmental indicators, 23, 29
   social vulnerability indicators, 36–39, 42–43, 50, 52
European Community Household Panel (ECHP), 39
Exposure, 18–20, 54–55, 57, 92, 115

G
Geographical Information System (GIS), 54, 94, 120
GINI coefficient, 23, 25
Governability, 15–16, 19–30, 55, 109, 111, 119–120, 138
   adaptation, 21
   community governability, 20–21
   definition, 15–16
   governability-environmental stress, 20, 26
   governability of Industrial Risk in Latin America, 23–29, 92
   market governability, 20
   state Governability, 20
Governance, 3–4
Grassroots movements, 93
Gross Domestic Product (GDP), 6, 24, 92, 119
Guatemala, 24–25, 44–49
   environmental indicators, 23, 29
   social vulnerability indicators, 36–39, 42–43, 50, 52

H
Hazardousness, 16–17, 20, 22, 94–99, 101–104, 106–113, 115, 118, 120–123, 127–129, 131, 133, 137–143
   assessment, 57, 71, 113, 137–139
   industries, 17–18, 95–96, 98–99, 109, 115, 118, 121–122, 134
Hazards
   industrial, 22, 29, 55
   natural, 55
Health, 36, 38–41, 47–49, 50–52
Honduras, 24–25, 44–49
   environmental indicators, 23, 29
   social vulnerability indicators, 36–39, 42–43, 50, 52
Hot-spots, 5, 57, 108, 127, 134, 137, 141, 144
Human Development Index (HDI), 6, 92

I
Iberian-Peninsula, 117
Ibero-America, 1, 5–13, 92, 139–140
   environmental Forum, 10–11
   environmental sustainability, 11
   GDP, 6–9, 23–25
   GDP and environmental deterioration, 23–25
   Human Development Index, 6
   Ibero-american summits, 6, 10–11
   identity, 10
   population, 8, 92, 96, 98–100, 102, 104
   regional integration, 10
   scientific cooperation, 11
   socio-economic variables, 6
   territory size, 7
Indicators, 22–23, 35–52
   approaches, 36–37
   Argentina, 23, 43
   Bahamas, 44–49
   Belize, 24, 44–49
   Bolivia, 24–25, 44–49
   Brazil, 24–25, 44–49
   Chile, 24–25, 44–49
   Colombia, 24–25, 44–49
   Costa Rica, 24–25, 44–49
   Cuba, 24–25, 44–49
   Dominican Rep, 24–25, 44–49
   Ecuador, 24–25, 44–49
   El Salvador, 24, 44–49
   environmental indicators, 23, 29
   Guatemala, 24–25, 44–49
   Guyana, 42, 44–49
   Haiti, 44–49
   Honduras, 24–25, 44–49
   Jamaica, 44–49
   Mexico, 24–25, 44–49

Indicators (*cont.*)
  Nicaragua, 24–25, 44–49
  Panama, 24, 44–49
  Paraguay, 24–25, 44–49
  Peru, 24, 44–49
  poverty indicators, 38
  risk indicators, 17
  selection criteria, 42
  social indicators, 36, 38–39
  social vulnerability index (SVI), 37
  socio-economic indicators, 39–41, 44
    access to drinking water, 123, 125–126
    annual proportion of deaths of children under 5 due to infectious intestinal diseases, 41
    basic needs, 100
    child death rate, 41, 47
    death rate of children under, 5, 41, 47
    death rate of women in childbirth, 41, 48
    definitive dependent population, 100, 123
    dwelling, 123–126
    health cover, 100, 123–124
    illiteracy in the population aged 15 to 24, 41
    incidence of tuberculosis, 41, 48
    life expectancy, 41
    literacy/education, 100, 123
    net rate of registration in primary education, 41, 45
    per capita Gross Domestic Product, 41, 44
    population with access to improved sanitary services, 41, 46
    population with sustainable access to supplies of drinking water, 41, 46
    profession/employment, 123
    proportion of doctors, 41, 49
    proportion of hospital beds, 41, 49
    proportion of pregnant women attended by qualified personnel, 41, 49
    proportion of under-weight newly-born children, 41, 49
    public health spending as a percentage of GDP, 41
    rate of dependence, 41, 44
    rate of literacy, 41, 45
    rate of use of contraceptive methods among women, 41, 48
    sewage services, 51, 100
    sewage treatment, 123
    single-parent homes, 51, 100
    single-parent households, 123, 126
    supply of drinking water, 46, 51, 100
    total fertility rate, 41, 48
    total spending on health per capita, 41
    transitory dependent population, 100, 123
    unemployment rate, 41
    urban population in situation of poverty, 41
    urban unemployment rate, 41, 44
    work/occupation, 100, 139
  Surinam, 44–49
  Trinidad & Tobago, 44–49
Indigence, 23, 36
  See also Poverty, extreme poverty
Industrial Hazardousness/perilousness, 57–65, 74–90, 137, 140
  aggregated, 57, 60, 64–66
  Bolivia, 69–90
  categories, 56, 61, 65
  levels, 55
  methodology of analysis, 57–65
  radius of influence (R), 58–61, 63
  Spain, 92
Inequality, 3–4, 18, 23, 25, 37, 140
ISO 14001, 23, 25, 28, 94

J
Justice, 3, 56, 91
  distributive justice, 3
  environmental justice, 56, 91

K
Kernel, 59–62
  density, 59–60
  function, 60, 62
Kuznet curve, theory, 3

L
Latin-America, 1, 140
  corporate social responsibility, 27
  environmental deterioration, 19, 23–25, 28
  GDP, 23–25
  GINI, 23–24
  inequality, 18, 23, 25, 35, 37
  ISO 14001 implementation, 23, 25, 28
  poverty, 18–19, 23–24, 28
Level of Environmental Complexity, 54, 61, 98
Living Conditions Survey, 40
Living Standards, 39–40, 51

Index

M
Methodologies, 53, 64–66, 115
   industrial hazardousness, 53, 57–65, 69, 74–79, 81–84, 87–89, 91, 94–97, 101–109, 111–113, 115, 117, 121–123, 127, 129, 131, 133–135
   paucity of data, 64–66
   risk assessment, 22–23
   social vulnerability, 35–52
Mexico, 24–25, 44–49, 92
   environmental indicators, 23, 29
   social vulnerability indicators, 44
Millenium Development Goals, 11, 38–39
Millenium Development Indicators, 29
Mining, 71–72, 93–94

N
Nearest k-th neighbour method, 63
Nicaragua, 24–25, 44–47, 49
   environmental indicators, 23, 29
   social vulnerability indicators, 44

O
Oruro, 70, 76–77, 79
Overseas investment, 9

P
Panama, 24, 44–49
   environmental indicators, 23, 29
   social vulnerability indicators, 44
Paraguay, 23, 29, 44–49
   environmental indicators, 23, 29
   social vulnerability indicators, 44
Partnership in Statistics for Development in the 21st Century (PARIS 21), 39
Perilousness, 17, 22
Peru, 24, 44–49
   environmental indicators, 23, 29
   social vulnerability indicators, 44
Pin Map, 57
Polluting industries, 58
Polluting particles, 95
Potosi, 70, 76–77, 79
Poverty, 1–5, 18–19, 23–25, 71
   Bolivia, 69–90
   definition, 15–16, 21
   extreme poverty, 23–24, 38, 92
      See also Indigence
   geographical distribution, 36–38
   indicators, 35–44, 50–51
   Latin America, 15, 19, 23–29
   poverty-environmental deterioration, 2–5
   poverty line (PL), 36
   poverty trap, 4
   poverty-vulnerability, 19
   rural poverty, 25, 36, 42, 46
   urban poverty, 23–25, 36

R
Raster, 57, 59
Resilience, 21–22
   definition, 15–16, 21
Rio de Janeiro, 10
Risk, 1, 3–6, 9–13, 15–30, 53–57, 60, 64–66, 69, 74–76, 79–81, 87–90, 117, 119–121, 123–124, 126–129, 132–135, 137–143
   assessment, 22–23, 53–57, 65–66
   Bolivia, 69–90
   cartography, 55–56
   definitions, 15
   dimensions, 35
   empirical assessment methodologies, 15, 17, 20, 26, 29–30
   evaluated risk, 4, 11–12, 15–17, 21–23, 25, 30, 120, 128, 132–134, 138, 142–143
   managed risk, 12, 21, 23, 55, 120, 135
   managers, 53
   potential risk, 1, 4–5, 15–16, 120, 129, 133–135, 142
   surfaces, 54, 58, 60–63

S
Santa Cruz, 69–71, 73–77, 79, 81–89
Single-parent families, 36
Small and medium enterprises (SMEs), 27–28
Social Capital, 21, 25
Social exclusion, 35, 37
Spain, 8–9, 92, 117–135, 138–141, 143
   Alicante, 124–127
   Aranda de Duero, 122
   Barcelona, 118, 121–125, 128–129
   census, 94–95, 99–100, 108, 111–115
   Elche, 125
   environmental indicators, 23, 29
   Gozón, 122
   Madrid, 117–118, 120, 123–125, 127–134
   Malaga, 122–126, 128
   Murcia, 122, 126–127, 129
   Prat de Llobregat (El), 122
   Rubí, 122
   Sant Cugat del Vallés, 122
   Saragossa, 121–122, 126–127, 129
   Seville, 120, 122, 124–126, 129–134

Spain (*cont.*)
  social vulnerability, 35–52, 120, 123–130, 133
  Spanish multinationals, 9, 118
  Valencia, 118, 122–123, 125–126, 128–129
  Valladolid, 122–123
  Vitoria-Gasteiz, 122
Spatial analysis, 57–58, 103
Stressors, 53–54
Sucre, 69–70, 73–77, 79, 81, 83–85, 87–89
Summits, 6, 10–11
  Guadalajara, 10
  Madrid, 10
  Salamanca, 11
Sustainability Science, 2–6
  definition, 2

T
Trinidad (Bol), 70

U
UN, 10, 38
  assistance framework for development, 38
  minimum national social data set, 39
  world Conferences, 10, 38
Uncertainty, 15–16, 20, 55
Unsatisfied basic needs (UBN), 36
Uruguay, 23–24, 44–49
  environmental indicators, 23, 29
  social vulnerability indicators, 36–39, 42–43, 50, 52

V
Venezuela, 24–25, 44–49
  environmental indicators, 23, 29
  social vulnerability indicators, 36–39, 42–43, 50, 52
Vicious circles, 2–5, 19
  definition, 2
  rural areas, 3
  urban areas, 3, 18
Vulnerability, 1, 3, 5–6, 9, 11–13, 16–22, 35–52, 74–76, 78–79, 81, 84–89, 92, 94, 96, 99–100, 104–109, 111–115, 120, 123–130, 133, 137–142, 144
  Argentina, 91–109
  Argentina, 91–115
  bio-physical vulnerability, 19
  Bolivia, 69–90
  Bolivia, 79–81
  definition, 16, 137
  intrinsic, 18
  relative, 19
  social vulnerability, 16–19, 22, 30, 35–52, 92, 94, 96, 99–100, 104–109, 111–114, 137–140, 142
  social vulnerability index (SVI), 37
  socio-economic vulnerability, 35–52
  Spain, 117–121, 123–128, 133, 135

W
Wealth-environmental deterioration, 2–4
  See also Economic Growth-Environmental deterioration